Push your Career Publish your Thesis

Science should be accessible to everybody. Share the knowledge, the ideas, and the passion about your research. Give your part of the infinite amount of scientific research possibilities a finite frame.

Publish your examination paper, diploma thesis, bachelor thesis, master thesis, dissertation, or habilitation treatises in form of a book.

A finite frame by infinite science.

An Imprint of
Infinite Science GmbH
MFC 1 | Technikzentrum Lübeck
BioMedTec Wissenschaftscampus
Maria-Goeppert-Straße 1
23562 Lübeck
book@infinite-science.de
www.infinite-science.de

Editor

Thorsten M. Buzug
Institute of Medical Engineering
University of Lübeck
buzug@imt.uni-luebeck.de

Reihe: Medizinische Ingenieurwissenschaft und Biomedizintechnik

Diese Reihe umfasst Werke der Medizinischen Ingenieurwissenschaft und Biomedizintechnik, deren Themen strategisch unter den Zukunftstechnologien mit hohem Innovationspotenzial anzusiedeln sind. Als wesentliche Trends dieser Forschungsgebiete, sind die Schlüsselbereiche Computerisierung, Miniaturisierung und Molekularisierung zu nennen. Bei der Computerisierung sind dabei die inhaltlichen Schwerpunkte beispielsweise in der Bildgebung und Bildverarbeitung gegeben. Die Miniaturisierung spielt unter anderem bei intelligenten Implantaten, der minimalinvasiven Chirurgie aber auch bei der Entwicklung von neuen nanostrukturierten Materialien eine wichtige Rolle, und die Molekularisierung ist in der regenerativen Medizin aber auch im Rahmen der sogenannten molekularen Bildgebung ein entscheidender Aspekt. Forschungs- und Entwicklungspotenzial werden auch der Biophotonik und der minimal-invasiven Chirurgie unter Berücksichtigung der Robotik und Navigation zugeschrieben. Querschnittstechnologien wie die Mikrosystemtechnik, optische Technologien, Softwaresysteme und Wissenstechnologien sind dabei von hohem Interesse.

Christian Kaethner

Elliptical Coils in Magnetic Particle Imaging

Medical Engineering Science and
Biomedical Engineering — Volume 6

Editor: Thorsten M. Buzug

© 2015 Infinite Science Publishing
the BioMedTec Science Campus Publisher Lübeck

An Imprint of Infinite Science GmbH,
MFC 1 | BioMedTec Wissenschaftscampus
Maria-Goeppert-Straße 1
23562 Lübeck

Cover Design, Illustration: Uli Schmidts, metonym
Copy Editing: University of Lübeck, Institute of Medical Engineering

Publisher: Infinite Science GmbH, Lübeck, www.infinite-science.de
Print: Books on Demand GmbH, Norderstedt

ISBN Paperback: 978-3-945954-07-2

Das Werk, einschließlich seiner Teile, ist urheberrechtlich geschützt. Jede Verwertung ist ohne Zustimmung des Verlages und des Autors unzulässig. Dies gilt insbesondere für die elektronische oder sonstige Vervielfältigung, Bearbeitung, Übersetzung, Mikroverfilmung, Verbreitung und öffentliche Zugänglichmachung sowie die Einspeicherung und Verarbeitung in elektronischen Systemen.

Die Wiedergabe von Gebrauchsnamen, Handelsnamen, Warenbezeichnungen usw. in dieser Publikation berechtigt auch ohne besondere Kennzeichnung nicht zu der Annahme, dass solche Namen im Sinne der Warenzeichen- und Markenschutz-Gesetzgebung als frei zu betrachten wären und daher von jedermann verwendet werden dürften.

Bibliografische Information der Deutschen Nationalbibliothek:
Die Deutsche Nationalbibliothek verzeichnet diese Publikation in der Deutschen Nationalbibliografie; detaillierte bibliografische Daten sind im Internet über http://dnb.d-nb.de abrufbar.

Bibliographic information published by the Deutsche Nationalbibliothek
The Deutsche Nationalbibliothek lists this publication in the Deutsche Nationalbibliografie; detailed bibliographic data are available in the internet at http://dnb.d-nb.de.

Abstract

Magnetic Particle Imaging is a functional imaging technique used to visualise the distribution of super-paramagnetic iron oxide nanoparticles by using magnetic fields. These fields are generated by current carrying coils specifically arranged in different scanner topologies. Up to now, the established geometries of a coil have been either circular shaped or rectangular shaped. The aim of this work is to evaluate the application of approximated elliptical coils in Magnetic Particle Imaging. Based on the clinical practice, it is conceivable to integrate a circular single-sided scanner, for example, into a patient table. In order to achieve an optimally adapted result to the medical application, a tradeoff between size, field of view and patient access must be made. The new coil geometry is designed to increase the field of view, without exceeding the width of the patient table and therefore ensure good access to the patient. Besides the integration of this new coil geometry into an existing simulation framework, the simulation is validated by comparing the results to a self-built coil and an already established coil geometry. In addition to this, the application of the coil in different scanner topologies is evaluated.

Kurzfassung

Magnetic Particle Imaging ist eine funktionelle Bildgebungsmodalität, welche eine Visualisierung der Verteilung superparamagnetischer Eisenoxid Nanopartikeln unter Ausnutzung von Magnetfeldern ermöglicht. Diese Magnetfelder werden durch stromdurchflossene Spulen generiert, die spezifisch in unterschiedlichen Scannertopologien angeordnet werden. Die heutzutage etablierten Spulengeometrien haben entweder eine kreisrunde oder eine rechteckige Form. Das Ziel dieser Arbeit ist zu evaluieren, ob sich eine Anwendung von approximierten elliptischen Spulen für Magnetic Particle Imaging eignet. Hinsichtlich des klinischen Alltags ist es denkbar, einen kreisförmigen single-sided Scanner in einen Patietentisch zu integrieren. Um ein möglichst optimal angepasstes Resultat hinsichtlich der medizinischen Applikation zu erzielen, muss ein Kompromiss zwischen Größe, Betrachtungsfeld und Patientenzugang gebildet werden. Die Idee der neuen Spulengeometrie ist, dass eine Vergrößerung des Betrachtungsfeldes erreicht werden kann, ohne die Breite des Patiententisches zu überschreiten und somit einen guten Patientenzugang zu gewährleisten. Neben der Integration dieser neuen Spulengeometrie in eine bestehende Simulationsumgebung wird die Simulation durch einen Vergleich mit einer selbstgebauten Spule und eine bereits etablierten Spulengeometrie validiert. Zusätzlich zu dieser Validierung wird die Anwendbarkeit der Spule für verschiedene Scannertopologien untersucht.

Contents

1	**Introduction**	**1**
2	**Theoretical Principles**	**5**
	2.1 Physical Basics	5
	2.1.1 Electrostatics	6
	2.1.2 Magnetostatics	8
	2.1.3 Maxwell Equations	10
	2.2 Magnetic Particle Imaging	13
	2.2.1 Super-paramagnetic Iron Oxide Nanoparticles	15
	2.2.2 Signal Encoding	19
	2.2.3 Spatial Encoding	21
	2.2.4 Reconstruction	23
	2.2.5 Coil Geometries	26
3	**Material and Methods**	**33**
	3.1 Simulation Framework	33
	3.2 Extension of the Simulation Framework	36
	3.2.1 Circle Segments	36
	3.2.2 Approximated Elliptical Coils	38
	3.3 Realisation of the Simulated Coil Geometry	41
	3.3.1 Geometry	41
	3.3.2 Construction of the Coil Mould	42
	3.4 Measurement Conditions	42
	3.5 Introduction of Error Metrics	46

4 Experiments and Results — 49
- 4.1 Simulations — 50
- 4.2 Measurements — 55
- 4.3 Comparison between Simulation and Measurement — 58
- 4.4 Circular Coils versus Approximated Elliptical Coils — 61
  - 4.4.1 Validation — 61
  - 4.4.2 Extension — 62
- 4.5 Application within a Circular Single-Sided Scanner — 65
  - 4.5.1 Validation — 65
  - 4.5.2 Extension — 67
- 4.6 Application within an Open MPI Scanner — 70
  - 4.6.1 Validation — 71
  - 4.6.2 Extension — 72

5 Discussion — 77
- 5.1 Comparison between Simulation and Measurement — 77
- 5.2 Validation — 78
- 5.3 Extension — 80

6 Summary and Outlook — 85

Bibliography

1
Introduction

Magnetic Particle Imaging (MPI) is a tomographic imaging modality for the detection of the spatial and temporal distribution of super-paramagnetic iron oxide (SPIO) nanoparticles. Thanks to their super-paramagnetic properties, the SPIO nanoparticles have an inherent non-linear magnetisation used for signal detection. To excite the nanoparticles, an oscillating magnetic field superimposed by a static magnetic field is used. The latter is a gradient field generating a field free point (FFP), in which the particle response can be used for imaging. The oscillating magnetic field enables a movement of this FFP through space. The main advantages of MPI are a high temporal and spatial resolution and a high sensitivity. In addition to this, MPI allows the acquisition of functional images without using radioactive contrast agents or tracers.

Different coil geometries can be used to generate a magnetic field. At the moment, there are two geometries established in MPI: circular shaped coils and rectangular shaped coils. Both coil geometries can be modified and adapted to the respective field of application. Important factors are, for example, the chosen dimensions of the coil, the currents necessary to generate the magnetic field and the coil arrangement. As a result of these factors, the field of view (FOV) generated varies. However, considering the medical applications, there are always advantages and disadvantages with respect to the

Chapter 1 | Introduction

coil geometry. This means the coil geometry and the corresponding arrangement of the coils should ideally be adapted to a medical application scenario.

The motivation of this work is based on the following simple application scenario in clinical practice: assuming there is a circular shaped single-sided MPI scanner integrated into a patient table, it is necessary due to the coil dimensions to have an idea about the medical application. To make an adequate decision regarding the coil dimensions, two aspects must taken be into account. First, the radius of the outer coil can be chosen to coincide with the width of the patient table. The advantage would be an excellent patient access, while the disadvantage would be a small FOV. Second, the radius of the coil can be increased to have a larger FOV, resulting in a more difficult access to the patient. The realisation of such a single-sided scanner based on circular coils leads to a compromise between the FOV, the coil dimensions and the patient access. The introduction of a new coil geometry could serve as an alternative solution to this problem. Based on a circular shaped coil, the new coil geometry must be modified to have a suitable width to be integrated into a patient table and have an appropriate size in terms of the FOV. A translation of this idea is conceivable by using elliptical coils. This may result in a extension of the FOV, while the width of the patient table could be maintained. As an adequate approximation to such a geometry, a coil based on two half circles connected by two straight parts is introduced.

The objective of this work is to evaluate the applicability of approximated elliptical coils for MPI. For this purpose, an existing framework to simulate the physical processes in MPI is extended with this new coil geometry. Two variations of D-shaped approximated elliptical coils are implemented in addition to the regular coil. To evaluate the simulation results, the newly introduced coil geometry is realised using a self-constructed coil mould. As a further validation, the simulated approximated elliptical coils are modified to the outer appearance of a circular shaped coil and compared to an existing circular shaped coil. Subsequently, the application is verified within a single-sided MPI scanner and an open MPI scanner. In this context, the scanners are rebuilt using approximated elliptical coils. The consideration of the applicability is then performed by validating the scanner using the new coil geometry and by extending the existing scanner geometry with certain modifications.

This thesis is structured in the following way: chapter 2 starts with an introduction of the most important physical concepts needed to understand the MPI principle. This includes a description of the theoretical basics, from the electromagnetic field theory to the Maxwell equations. Further, this chapter gives an insight into the MPI principle itself.

In chapter 3, the simulation framework used is introduced and a description is provided of how the simulation framework is extended with a new feature and a new coil geometry. In addition to the integration of a new coil geometry, the realisation process and the corresponding measurement conditions are described. For an objective comparison of the results, a number of selected error metrics are introduced. Chapter 4 deals with the experiments carried out and the corresponding results to validate and extend the usability of the new coil geometry. This resulting knowledge is afterwards discussed in chapter 5. Finally, in chapter 6 the thesis is summarised and some prospective research possibilities are given.

2
Theoretical Principles

The aim of the following chapter is to impart a basic knowledge of the fundamental principles that are important to this work. The first part provides an introduction to electromagnetic field theory (section 2.1). This chapter also provides an introduction to MPI (section 2.2). The purpose of these sections is to give an idea of how MPI works, how a signal can be produced and how it is possible to use this signal to reconstruct an image. At this stage, the concept is mainly described for one-dimensional scenarios, however an extension to two-dimensional and three-dimensional scenarios is perfectly possible.

2.1 Physical Basics

With respect to the physical aspects relevant to MPI, the following sections outline the principles of electrodynamics. Electrodynamics is a branch of physics that deals with electrostatics and magnetostatics, namely electric fields and magnetic fields. The main source for this section and in particular the formulas used here, is [46].

Chapter 2 | Theoretical Principles

2.1.1 Electrostatics

The basic concept of electrostatics is the consideration of the phenomenon of electrification. In this context the force $\vec{F}$, proportional to a given charge Q, serves as a measure of the strength of electrification. In general, the force between two charged objects can be described by the fact that equally charged objects repel each other, while those with an opposite sign attract each other. Considering two charged sources Q_i and Q_j with a distance r, the force acting on the first source can be written as

$$\vec{F}_i \sim \frac{Q_i \cdot Q_j}{r^2} \cdot \vec{e}_{ji}, \tag{2.1}$$

where $\vec{e}_{ji}$ is a unit vector pointing from the j-th to the i-th object. Replacing the proportionality with an equals sign gives a definition of the unit of charge. Expanding the equation with a factor depending on the surrounding medium, including the permittivity ε_0, the charge can be calculated with

$$\vec{F}_i = \frac{1}{4\pi\varepsilon_0} \frac{Q_i \cdot Q_j}{r^2} \cdot \vec{e}_{ji}, \tag{2.2}$$

which is known as Coulomb's law or Coulomb's inverse-square law [45]. In general, the factor ε_0 is replaced by ε with a specific value for each surrounding medium. For air, this factor is given by $\varepsilon_0 \approx 8.854 \cdot 10^{-12} \mathrm{A\,s\,V^{-1}\,m^{-1}}$.

Coulomb's law describes the behaviour of two charged objects towards each other. For a system with more than two charged objects, the law can be extended by the superposition principle. Assuming a system with $N+1$ charged objects, the force acting on the i-th object is given by

$$\vec{F}_i = \frac{1}{4\pi\varepsilon_0} \sum_{j=1}^{N} \frac{Q_i \cdot Q_j}{|\vec{r}_i - \vec{r}_j|^2} \frac{\vec{r}_i - \vec{r}_j}{|\vec{r}_i - \vec{r}_j|} = \frac{Q_i}{4\pi\varepsilon_0} \sum_{j=1}^{N} \frac{Q_j \cdot (\vec{r}_i - \vec{r}_j)}{|\vec{r}_i - \vec{r}_j|^3}. \tag{2.3}$$

For a continuous charge density $\varrho(\vec{r}')$ the equation can be reformulated to

$$\vec{F} = \frac{Q}{4\pi\varepsilon_0} \iiint_{V'} \frac{\varrho(\vec{r}') \cdot (\vec{r} - \vec{r}')}{|\vec{r} - \vec{r}'|^3} \, \mathrm{d}V', \tag{2.4}$$

where the considered volume V' describes the region in which the charge density is not equal to zero.

An extension of the previous considerations is the introduction of an electric field $\vec{E}(\vec{r})$

at a point charge or an arrangement of charges. It is defined as the force acting on a unit charge q at a specific location

$$\vec{E}(\vec{r}) := \frac{\vec{F}}{q}. \qquad (2.5)$$

As an analogy to equation (2.3) and (2.4), the formulation for the electric field can be written as

$$\vec{E}(\vec{r}) = \frac{1}{4\pi\varepsilon_0} \sum_{j=1}^{N} \frac{Q_j \cdot (\vec{r} - \vec{r}_j)}{|\vec{r} - \vec{r}_j|^3}, \qquad (2.6)$$

whereby Q_i can be replaced by the unit charge q. Accordingly, for a continuous charge density, the electric field is established by

$$\vec{E}(\vec{r}) = \frac{1}{4\pi\varepsilon_0} \iiint_{V'} \frac{\varrho(\vec{r}') \cdot (\vec{r} - \vec{r}')}{|\vec{r} - \vec{r}'|^3} \, dV'. \qquad (2.7)$$

The previous assumptions relate only to an object located in a vacuum. In reality, the presence of matter in the surrounding medium has an important role. Generating an electric field is not easily possible in every case. However, at this point reference is made to [45, 46]. With regard to the circumstances concerning this work, air is always the surrounding medium and therefore an electric field can be generated.

If one now considers an electrical conductor, it is apparent that the charges are freely displaceable. This implies that a charge in a conductor moves until the electric field inside the conductor disappears and the potential

$$\varphi(\vec{r}) = \frac{1}{4\pi\varepsilon_0} \iiint_{V'} \frac{\varrho(\vec{r}')}{|\vec{r} - \vec{r}'|} \, dV' \qquad (2.8)$$

becomes constant [46]. This fact leads to a direct relationship between the electric field and the potential:

$$\varphi(\vec{r}) = -\int_{\vec{r}_a}^{\vec{r}} \vec{E} \, d\vec{l}. \qquad (2.9)$$

It should be mentioned that the potential in $\vec{r}_a$ is normalised to zero.

Regarding the Maxwell equations in section 2.1.3, the electric displacement field $\vec{D}$ is briefly introduced at this point. It is used for the characterisation of the electric field lines in a certain volume that is full of material. With respect to the 'free' charge density

ϱ, the electric displacement can be calculated by

$$\oiint_{\partial V} \vec{D} \, \mathrm{d}\vec{F} = \iiint_V \varrho \, \mathrm{d}V. \tag{2.10}$$

Another variable which should be mentioned in this context is the current density $\vec{J}$. Due to the law of charge conservation [45], the current flow must disappear through a closed surface ∂V:

$$\oiint_{\partial V} \vec{J} \, \mathrm{d}\vec{F} = 0. \tag{2.11}$$

As described in [46], the charges would otherwise accumulate, which implies time-varying electric fields.

2.1.2 Magnetostatics

Magnetostatics can be described as the magnetic analogue of the electrostatics. In contrast to static electricity, which considers steady electric currents, magnetostatics deals with temporally constant magnetic currents. Furthermore, the appropriate field theory is an important part of the Maxwell equations described in section 2.1.3. With regard to magnetisation, it is generally possible to distinguish between permanently active, temporarily active and non-magnetic materials. It can be observed that each magnet has a defined direction. The currents are particularly pronounced at the respective ends of a magnet. A distinction can be made on the basis of the effect, since the magnetisation of the so-called north pole and south pole differs in its sign.

In analogy to Coulomb's law in electrostatics, introduced by equation (2.2), the formulation in magnetostatics is

$$\vec{F}_i = \frac{1}{4\pi\mu_0} \frac{p_i \cdot p_j}{r^2} \cdot \vec{e}_{ji}. \tag{2.12}$$

Here, $\mu_0 = 4\pi \cdot 10^{-7} \mathrm{V\,s\,A^{-1}\,m^{-1}}$ is the magnetic permeability for a vacuum and p_i as well as p_j denote the magnetic poles. It should be noted that the use of a defined point for the pole, similar to the point charge of electrostatics, is a conceptual model.

A magnetic force acting on two magnetic poles is defined by equation (2.2). For a description of a force acting on several independent magnetic monopoles, as in electrostatics, the superposition principle can be used. Analogous to equation (2.3) is

$$\vec{F} = \frac{p}{4\pi\mu_0} \sum_{j=1}^{N} \frac{p_j \cdot (\vec{r}-\vec{r}_j)}{|\vec{r}-\vec{r}_j|^3}, \qquad (2.13)$$

wherein it should be noted that this represents a simplification due to the absence of a magnetic monopole. Equal to the definition of the electric field, the simplified pole p is omitted to describe the magnetic field strength as

$$\vec{H}(\vec{r}) := \frac{\vec{F}}{p} \qquad (2.14)$$

or with respect to equation (2.6) it can be defined by

$$\vec{H}(\vec{r}) = \frac{1}{4\pi\mu_0} \sum_{j=1}^{N} \frac{p_j \cdot (\vec{r}-\vec{r}_j)}{|\vec{r}-\vec{r}_j|^3}. \qquad (2.15)$$

Two additional variables, essential for the understanding of the Maxwell equations in section 2.1.3 and the theoretical background of this work, are the magnetic field $\vec{B}$ and the magnetisation $\vec{M}$. The latter is used for the characterisation of a magnetic material, for which purpose an observed volume V is set in relation to a magnetic moment $\vec{m}$. The magnetisation is calculated with

$$\vec{M} = \lim_{\Delta V \to 0} \frac{1}{\Delta V} \sum_j \vec{m}_j. \qquad (2.16)$$

For instances where there is non-homogeneous magnetisation equivalent to the electric charge density, a magnetic charge density

$$\varrho_m = -\mu_0 \, \text{div} \, \vec{M} \qquad (2.17)$$

is defined. By combining the magnetic field strength $\vec{H}$ and the magnetisation $\vec{M}$ it is possible to calculate the magnetic field $\vec{B}$. The equation reads

$$\vec{B} = \mu_0 \left(\vec{H} + \vec{M} \right), \qquad (2.18)$$

wherein the unit of this size is Tesla $T = V\,s\,m^{-1}$.

One of the most important connections between electricity and magnetism is concerned with the feasibility of using current-carrying coils in MPI. It deals with the effect of the magnetic influence of the electric current. A calculation of the resulting magnetic field can be carried out using Ampère's circuital law or by applying the law of Biot-Savart. For this work, a detailed illustration of the former is omitted, since application is limited to the calculation of magnetic fields for simple symmetrical currents. For a more detailed investigation, reference can be made at this point to [14]. However, it should be noted that this law is a part of the Maxwell equations and is therefore briefly described in section 2.1.3.

The law of Biot-Savart states that a current I along an arbitrary path can be considered as a composition of flows of many infinitesimal current components. For the purposes of calculation, a distinction needs to be made between thin current-carrying conductors S and volume elements dV'. The former applies to

$$\vec{H}(\vec{r}) = \frac{I}{4\pi} \oint_S \frac{d\vec{l'} \times (\vec{r} - \vec{r'})}{|\vec{r} - \vec{r'}|^3}, \tag{2.19}$$

whereby the magnetic effect of a volume element can be calculated via

$$\vec{H}(\vec{r}) = \frac{1}{4\pi} \iiint_{V'} \frac{\vec{J}(\vec{r'}) \times (\vec{r} - \vec{r'})}{|\vec{r} - \vec{r'}|^3} dV'. \tag{2.20}$$

2.1.3 Maxwell Equations

The preceding sections have provided a basic understanding of some of the key concepts of electrostatics and magnetostatics. Here, both areas of physics were treated independently as far as possible. It became clear that charges lead to electric fields, whereas currents are the cause of magnetic fields. The introduction of the Maxwell equations by J. C. Maxwell in the years 1861 to 1864 connected these two aspects [50, 51, 52].

To set up the Maxwell equations, the field concept of M. Faraday is used as described, for example, in [45, 46]. A simple description of his experiments with regard to electricity and magnetism or rather the corresponding fields served Faraday as motivation to introduce the concept of field lines.

In general, it can be said that, by using Maxwell's equations, a relationship can be defined between the following physical variables:

- electric field strength $\vec{E}$, in $V\,m^{-1}$,

- electric displacement field $\vec{D}$, in $A\,s\,m^{-2}$,
- magnetic field $\vec{B}$, in $V\,s\,m^{-2}$,
- magnetic field strength $\vec{H}$, in $A\,m = T$,
- current density $\vec{J}$, in $A\,m^{-2}$,
- electric charge density ϱ, in $A\,s\,m^{-3}$.

It should be noted that the physical variables listed here are exclusively related to the electromagnetic theory.

Combining the physical variables mentioned above leads to the fundamental laws in electromagnetic theory. It is thanks to J. C. Maxwell that a relationship between these individual conditions could be established. He postulated a relationship between the field of magnetism and electricity with the introduction of the displacement current density $d\vec{D}/dt$ equivalent to the current density $\vec{J}$. In summary, this yielded the following laws given in integral form:

$$\oint_{\partial F} \vec{E}\,d\vec{l} = -\frac{d}{dt}\iint_F \vec{B}\,d\vec{F}, \tag{2.21}$$

$$\oint_{\partial F} \vec{H}\,d\vec{l} = \iint \vec{J}\,d\vec{F} + \frac{d}{dt}\iint_F \vec{D}\,d\vec{F}, \tag{2.22}$$

$$\oiint_{\partial V} \vec{D}\,d\vec{F} = \iiint_V \varrho\,dV, \tag{2.23}$$

$$\oiint_{\partial V} \vec{B}\,d\vec{F} = 0. \tag{2.24}$$

Faraday's law of induction given by (2.21) describes the induction of an electric field by a time-varying magnetic field. Equation (2.22) is known as Ampère's circuital law. This law can be described or calculated in two different ways. The first one states that a magnetic field strength can be generated by an electrical current as declared in the original description of this law. The other one is given by Maxwell, who says that such magnetic fields can also be generated by changing electric fields. Equation (2.23) is Gauss's law. This characterises an electric displacement field relative to an electric charge density. It implies that, in an electric field with a positive and a negative charge, the field lines have their origins in these charges, so the direction of the field lines lead from the positive to the negative charge. The last equation connected with the Maxwell equations is Gauss's law for magnetism given by (2.24). It states that the absolute magnetic flux through a closed surface, called the Gaussian surface, is zero. Or to rephrase, it can be said that the generated magnetic field is a solenoidal vector field [30].

The application of the equations derived from Maxwell leads to the following situation: for an analysis based on integrals, the equations shall apply for any area or volume. The problem resulting from this is that for each application the determination of all components of a field, the respective volume must be selected. Additionally, these values may change and thus different volumes may be chosen. Therefore, for reasons of practicality, a transition to infinitesimally small areas or volumes is necessary. This leads to the following relationships as mentioned in [30, 46], which enable the transition from integral to differential notation:

$$\int_L \text{grad } s \, d\vec{l} = s\big|_{\vec{r}_a}^{\vec{r}_e}, \tag{2.25}$$

$$\iint_F \text{rot } \vec{v} \, d\vec{F} = \oint_{\partial F} \vec{v} \, d\vec{l}, \tag{2.26}$$

$$\iiint_V \text{div } \vec{a} \, dV = \oiint_{\partial V} \vec{a} \, d\vec{F}, \tag{2.27}$$

with L being a line with starting position $\vec{r}_a$ and ending position $\vec{r}_e$. Equations (2.26) and (2.27) are referred to as Stokes' and Gauss's theorem. Besides the importance in physics, for example in terms of Maxwell's equations, both are essential components of mathematics in the field of vector analysis.

The introduced differential equations grad, rot and div in coordinate representation can be written as

$$\text{grad } s = \frac{\partial s}{\partial x}\vec{e}_x + \frac{\partial s}{\partial y}\vec{e}_y + \frac{\partial s}{\partial z}\vec{e}_z, \tag{2.28}$$

$$\text{rot } \vec{v} = \left(\frac{\partial v_z}{\partial y} - \frac{\partial v_y}{\partial z}\right)\vec{e}_x + \left(\frac{\partial v_x}{\partial z} - \frac{\partial v_z}{\partial x}\right)\vec{e}_y + \left(\frac{\partial v_y}{\partial x} - \frac{\partial v_x}{\partial y}\right)\vec{e}_z, \tag{2.29}$$

$$\text{div } \vec{v} = \frac{\partial v_x}{\partial x} + \frac{\partial v_y}{\partial y} + \frac{\partial v_z}{\partial z}. \tag{2.30}$$

Expressing Maxwell's equations in differential form results in another important advantage over the integral notation. Indeed the equations (2.21) to (2.24) represent the respective field variables in relation to each other, but unlike the differential notation, an explicit solution to one of the variables used is not possible. By combining the Maxwell equations in integral form and the connections mentioned in the equations (2.25), (2.26) and (2.27), it is possible to produce the expression in differential form:

$$\text{rot } \vec{E}(\vec{r},t) = -\frac{\partial}{\partial t}\vec{B}(\vec{r},t), \tag{2.31}$$

$$\text{rot } \vec{H}(\vec{r},t) = \vec{J}(\vec{r},t) + \frac{\partial}{\partial t}\vec{D}(\vec{r},t), \tag{2.32}$$

$$\text{div } \vec{D}(\vec{r},t) = \varrho(\vec{r},t), \tag{2.33}$$

$$\text{div } \vec{B}(\vec{r},t) = 0. \tag{2.34}$$

2.2 Magnetic Particle Imaging

In modern medicine, the use of various imaging methods has become established as a helpful basis for decision-making in diagnostics. Depending on the medical application, different tomographic techniques can be used to visualise the internal structures of the human body. In contrast to, for example, conventional X-ray images, the images in computed tomography (CT) are free from superpositions. Furthermore, a distinction is made between morphological and functional imaging techniques. The former enables physical parameters to be measured in a direct way. These can include the attenuation of X-rays, as is the case in CT [6], or the determination of the proton density of hydrogen atoms, which occurs in magnetic resonance imaging (MRI) [48]. In both cases the modality depicts the anatomical structures and tissues of organs with a high resolution. One possible way to extend the applicability of these imaging techniques is to use radiocontrast agents [4, 42, 56, 77], which are a kind of medical medium designed to improve the contrast of specific internal body structures. Such contrast agents are injected or given orally to accumulate specifically in a desired region or tissue. Afterwards the spatial distribution of the used material is measured. Contrast agents for CT are typically iodine or gold-based nanoparticles [2, 29]; for MRI there are for example nanoparticles with a core of Gd^{3+}, Fe^{3+} or Mn^{3+} [40]. This improvement with contrast agents correlates to the other branch of the imaging techniques outlined above: functional imaging. These methods are a way of representing the interior of the human body indirectly, including imaging modalities such as positron emission tomography (PET) and single-photon emission tomography (SPECT) where radioactive tracer material is administered [76].

In 2005 a new approach in functional imaging known as Magnetic Particle Imaging (MPI) was published by B. Gleich and J. Weizenecker [15]. Similar to PET and SPECT, MPI visualises morphological processes, but instead of using radioactive tracer material, MPI uses super-paramagnetic iron oxide (SPIO) nanoparticles. Responding to a magnetic field, the distribution of these particles can be used to characterise a certain region of interest (ROI). Compared to conventional imaging techniques, MPI has certain advantages that are exemplified in Figure 2.1. In this case it is limited to the presentation of the most distinguishing characteristics: spatial resolution, temporal resolution and sensitivity. As shown, each of the chosen conventional techniques has certain advantages and disadvantages, whereas MPI promises good results in all three cases. In addition, it is not necessary to use ionising radiation as in CT. If the applied magnetic fields are used in an acceptable range for humans, MPI is a very gentle process that offers tremendous potential.

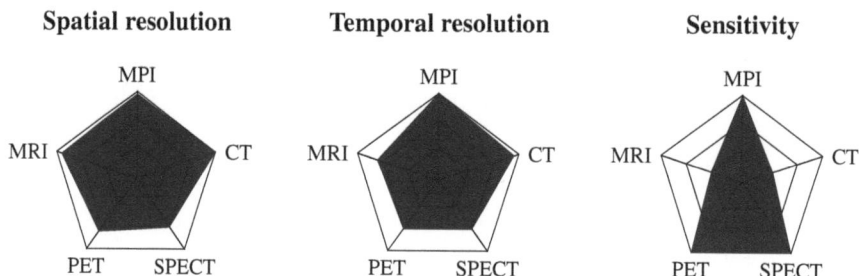

Figure 2.1: Comparison of MPI with other imaging methods with respect to spatial resolution, temporal resolution and sensitivity.

The combination of a high spatial and temporal resolution, a high sensitivity and the fact that MPI is able to acquire real-time images is the major advantage of this new imaging modality. As described in [3, 32], there is already a certain number of potential applications. One of these is the sentinel lymph node biopsy for breast cancer [12, 21, 41, 63]. The nature of breast cancer means that there is a possibility that the tumour will form metastases. These metastases can be detected by dissecting the sentinel lymph nodes for a histological examination. At the present time, surgical intervention strategies to locate the sentinel lymph nodes are based on scintigraphy and blue dye. Despite the fact that both methods achieve good results, MPI promises more precise surgical planning and more gentle treatment for the patient.

Achieving fast dynamic imaging with MPI, blood flow diagnosis of coronary heart diseases is another interesting possibility. The first in-vivo real-time images of a beating mouse heart were published in [75]. Examining a volume of 20.4 mm × 12 mm × 16.8 mm with a temporal resolution of 21.5 ms, all four chambers of the heart were visualised. An additional application in the context of the blood flow has been shown in [61]. This work discusses the feasibility of in-vivo recordings of cerebral blood flow in the brain of a mouse. This application could be, for example for the diagnosis of stroke, an alternative to conventional methods [62], such as Doppler / duplex ultrasound sonography, angiography, CT or MRI.

2.2.1 Super-paramagnetic Iron Oxide Nanoparticles

MPI is an imaging technique that relies on the determination of the spatial distribution of SPIO nanoparticles injected into the human body. An important point regarding such particles is that a distinction has to be made between contrast agents and tracers. Applying contrast agents, such as in CT or MRI, improves the tissue contrast and has advantages in separating similar tissues such as muscles and blood vessels. Tracers on the other hand used in PET, SPECT or MPI enable processes to be visualised that would otherwise not be apparent. Examples of such processes are the detection of tumour cells or the focus of inflammations [49].

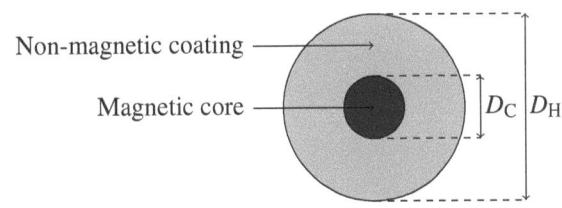

Figure 2.2: The left side shows magnetic nanoparticles in fluid form exposed to a permanent magnet. As shown in the schematic illustration on the right, each of the nanoparticles consists of a magnetic core and a non-magnetic coating.

The size of the nearly spherical tracers varies between 40 nm to 400 nm, whereby it is important to distinguish between the hydrodynamic diameter D_H of a particle and the core diameter D_C. As shown on the right side in Figure 2.2, the particle structure exhibits a magnetic core and a non-magnetic coating. The coating or the thickness of the coating is an important factor with regard to biological compatibility and potential in medical applications [40]. In addition to this, it is responsible for preventing agglomeration between the particles. The material of the coating could for example be dextran or carboxydextran [43]. Despite the significance of the non-magnetic coating, for this work the core is of higher importance due to its magnetic properties. Typically the size of the core is about 1 nm to 30 nm. Consisting for example of magnetite or maghemite it is responsible for the nanoparticle's magnetic behaviour [1]. An exemplary illustration of this fact is shown on the left side in Figure 2.2, where particles in fluid form are magnetised with a permanent magnet and are attracted to it.

With regard to magnetisation, one key step is to differentiate between the type of magnetisation. Here, a distinction has to be made between diamagnetism, ferromagnetism and paramagnetism [8]. The inner magnetic field of diamagnetic material decreases proportionally to the field strength of an external magnetic field. A ferromagnetic material has an own magnetisation. Bringing it into an external magnetic field, only the direction but not the amount is affected. The core of the magnetic nanoparticles is made of a ferromagnetic material, nevertheless they are exhibiting a super-paramagnetic behaviour.

In paramagnetic materials, a permanent magnetic dipole moment exists. If no external magnetic field is applied, the orientation of the dipole moments is arbitrary. There are no observable magnetic effects. If the particles are brought into an external magnetic field, for example a magnetic field generated by a coil, a rotation impulse is generated, so that the dipole moments of the magnetic core align parallel to the magnetic field. Switching the external magnetic field off results in a destruction of the alignment of the dipoles due to thermal fluctuations. In terms of the size of the particles and the thickness of the shell, no interactions occur between the particles. These and the high magnetic moment are the reason for the affix 'super' [31].

With respect to several anisotropic factors, in reality the magnetic moments depend on the magnetic field strength. Therefore, the magnetic moment of the SPIO nanoparticles prefers certain directions, introducing relaxation effects. On the one hand, there is the Néel relaxation, which describes the orientation of the particles relative to a changing magnetisation [57, 58]. On the other hand, there is the Brown relaxation describing the independent geometrical orientation of a nanoparticle [5].

As described in [33] it is not possible to determine the precise position of single nanopar-

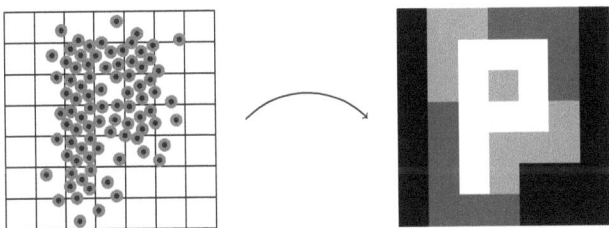

Figure 2.3: The left side shows an exemplary distribution of nanoparticles in a given small volume. By way of comparison, the corresponding grey-value image is shown on the right (adapted from [33]).

ticles in MPI. Instead of this, the concentration

$$c = \frac{N_P}{\Delta V} \quad (2.35)$$

of a certain number of particles N_P in a small volume ΔV is visualised (see Figure 2.3). Besides determining the spatial distribution of the SPIO nanoparticles, the calculation of magnetisation is the important factor in MPI.

If an external magnetic field is applied, the magnetisation course of the particles describes a non-linear curve. Calculating the magnetisation, it must be considered that each particle has its own magnetic moment m, which is randomly distributed. Thereby the magnetic moment can be written as

$$m = \frac{1}{6}\pi D_C M_S, \quad (2.36)$$

where M_S is the saturation magnetisation of the particles. The saturation magnetisation is reached when the magnetisation curve flattens and becomes nearly constant. The reason for this is that maximum magnetisation is reached and most of the particles align with the magnetic field strength H. In [33] it is stated that this is described as saturation field strength. It is also said that this region is reached for a magnetisation whose value is about 80% of the saturation magnetisation. Calculating the particle magnetisation relies on the Langevin theory [7], so the absolute value can be described by

$$M(H) = \begin{cases} cm\left(\coth(\xi) - \frac{1}{\xi}\right) & \text{for } H > 0 \\ 0 & \text{for } H = 0 \end{cases}, \quad (2.37)$$

Chapter 2 | Theoretical Principles

whereby

$$\xi = \frac{\mu_0 m H}{k_B T}. \qquad (2.38)$$

Here, μ_0 denotes the permeability of the surrounding space, $k_B = 1.380\,648\,8\,\text{J}\,\text{K}^{-1}$ is the Boltzmann constant and T denotes the temperature of the particles. To illustrate the influence of different core diameters on the magnetisation curve, Figure 2.4 can be considered. The larger the core diameter the higher the spatial resolution and the gradient (see Section 2.2.3). Therefore, the ideal magnetisation curve for MPI will be a step-function as shown as the black dashed line in Figure 2.4. However, it should be noted that unlimited extension of the particle size is not possible due to the fact that the super-paramagnetic properties of the particles will disappear above a certain size, limitations in relaxation effects and the medical applicability.

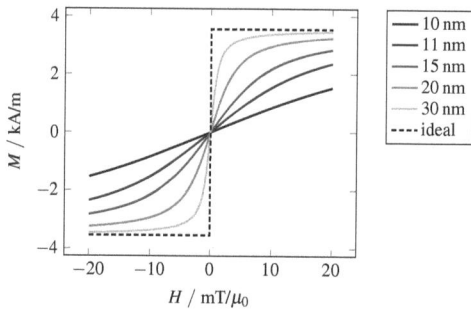

Figure 2.4: Visualisation of the super-paramagnetic behaviour of the SPIO nanoparticles for different core diameters (green). Additionally, the ideal particle magnetisation is displayed as a black dashed line.

Another possibility for deriving a statement regarding the resolution is to consider the derivation of the Langevin function and therefore the characteristics of the magnetisation. As described in [33] the derivative is given by

$$M'(H) = \begin{cases} cm\left(\frac{1}{\xi^2} - \frac{1}{\sinh^2(\xi)}\right) & \text{for } H > 0 \\ \frac{1}{3} & \text{for } H = 0 \end{cases}. \qquad (2.39)$$

The resulting curve has its maximum for $\xi = 0$ and becomes nearly zero reaching the saturation region (see Figure 2.5). To describe the resolution or the width of the area in which the magnetisation changes rapidly, the full width at half maximum (FWHM) can be used. It should be noted that, in this context, the resolution is the possibility to resolve two separate points [10]. To calculate the FHWM, the value M_{max} must be determined.

Subtracting the corresponding x-axis values of the value $\frac{M_{max}}{2}$ from each other yields the FHWM. As shown in Figure 2.5, the FHWM covers the area in which the magnetisation changes rapidly quite well. The FHWM depends very much on the gradient strength and slope of the magnetisation.

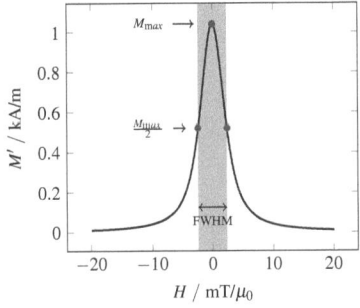
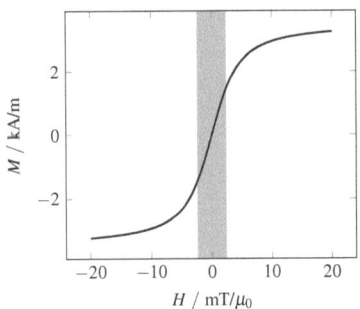

Figure 2.5: On the left the derivative of the Langevin function (green) with the calculated FWHM (grey area) is shown. The corresponding Langevin function as well as the FWHM can be seen on the right.

2.2.2 Signal Encoding

Signal encoding in MPI relies on the magnetic behaviour of the nanoparticles used. As described in section 2.2.1, applying an external magnetic field causes magnetisation $M(\vec{r},t)$ of the particles. The most important part of this is that the resulting magnetisation curve is non-linear. Therefore, it is possible to detect an induced electrical voltage

$$u(t) = -\mu_0 \frac{d}{dt} \int_\Omega M(\vec{r},t) h(\vec{r}) \, d^3\vec{r}, \quad (2.40)$$

which can ultimately be used to decode the signal from the particles. This formula, calculating the induced voltage $u(t)$ by setting the temporal change of the magnetisation $M(\vec{r},t)$ in relation to the sensitivity

$$h(\vec{r}) = \frac{1}{4\pi} \iiint_{V'} \frac{\vec{J}(\vec{r}') \times (\vec{r} - \vec{r}')}{|\vec{r} - \vec{r}'|^3} \, dV', \quad (2.41)$$

is known as the law of reciprocity. The sensitivity refers to the analysis of the field-generating properties of an electromagnetic coil using a unit current $I = 1\,\text{A}$ [32].

A sinusoidal magnetic field is usually used to excite the particles. This field is called either an excitation field or drive field, depending on whether the excitation of the par-

Chapter 2 | Theoretical Principles

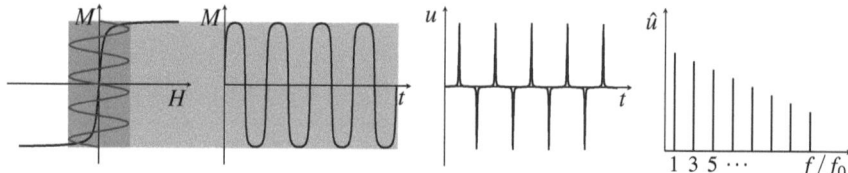

Figure 2.6: Basic procedure for signal encoding of MPI. In the first step, the particles are excited with a sinusoidal magnetisation curve. The magnetisation of the particles is non-linear. The resulting magnetisation induces a voltage. The Fourier transformation of the voltage contains not only the fundamental frequency but also harmonics.

ticles should be in the foreground or the movement of a field-free point (FFP) through the spatial domain, as described in section 2.2.3. The calculation is performed according to

$$H_D(t) = H_0 \cdot \sin(2\pi f_0 t), \qquad (2.42)$$

where H_0 is the amplitude of the magnetic field, f_0 is the excitation frequency and t is the time. An example of exciting the particles with such a field is shown in Figure 2.6. In this case, the non-linear magnetisation causes a nearly rectangular distortion of the sinusoidal magnetisation curve. As mentioned above, an electrical voltage is induced which is proportional to the negative time derivative of the magnetisation. It should be noted that not only the magnetised particles induce a voltage but also the excitation field itself. However, the resulting voltage signal is several orders of ten higher than the signal of the particles. Therefore, in order to use the signal of the particles for imaging, it is necessary to separate the signals using filtering. This filtering is possible, since the induced signal of the excitation field contains only one frequency.

Alternative ways of creating magnetic fields have been proposed in [60]. In contrast to the non-linear magnetisation of the nanoparticles, no higher harmonics can be generated by a linear magnetisation curve. An example of this is shown in Figure 2.7. The Fourier transformation of the particle response shows only the first harmonic, which refers to the voltage induced by the excitation field.

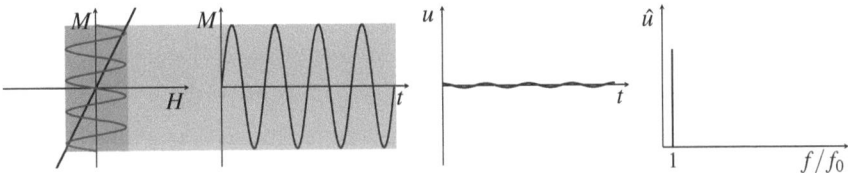

Figure 2.7: In cases where the magnetisation curve is linear, the induced voltage will be nearly zero. Other than the first harmonic, none of the higher harmonics can be measured.

2.2.3 Spatial Encoding

To characterise the distribution of the SPIO nanoparticles, not only are the concentration values needed but so too is the spatial position in the ROI. Here, the drive field is superimposed upon with a static magnetic field $H_S(t) = Gx$, which is called a selection field. The parameter G describes the gradient strength, which increases linearly between $[-H_0/G, H_0/G]$. At a certain point of the selection field, called FFP, the magnetic field is zero. Only at this point, or very near to it, can a change in the magnetisation of the SPIO nanoparticles be measured. At all other points in space, the magnetisation of the nanoparticles is constant, so that these points are not contributing to the measurement signal. As well as the concept of the FFP, a field-free line (FFL) can be used to encode the spatial position of the SPIO nanoparticles. This concept is characterised by a higher sensitivity, especially for three-dimensional imaging, and was first published in [74]. The implementation of the FFL concept was questionable due to a required power that was a thousand times higher than the FFP concept. In [36, 38] it has been shown that this disadvantage can be compensated. However, it should be noted that the FFL concept requires further assumptions and that this work is focused on the principle of the FFP.

By superimposing of the drive field and selection field, the overall magnetisation consequently changes. It follows

$$H(x,t) = H_D(x) + H_S(t) = H_0 \cdot \sin(2\pi f_0 t) + Gx. \tag{2.43}$$

The movement of the FFP is thereby restricted to the above mentioned range of the gradient field $[-H_0/G, H_0/G]$. As described in [32] for a step-function as particle response, the change of the magnetisation can only be measured at $x = -\frac{H_0}{G}\sin(2\pi f_0 t)$. That means that there is a direct connection between the measured signal and the spatial distribution of the particles. However, in reality the particle response is equivalent to the Langevin function (see equation (2.37) and Figure 2.4). This means that the range for imaging is limited by the gradient. The higher the slope of the gradient the smaller the area for imaging and vice versa. Additionally, the slope of the gradient is adapted to the image resolution. Outside this dynamic range, the particles are in saturation. This situation is illustrated, for example, in Figure 2.8. For particles in the saturation area there is only a small magnetic response and therefore the amount of induced current decreases as well. The resulting frequency spectrum contains only the first harmonic of the excitation frequency and a few higher harmonics, which can hardly be measured.

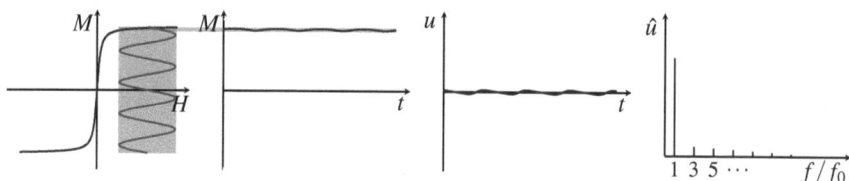

Figure 2.8: Overlapping of the drive field with a static selection field. The particles are saturated, so the magnetisation is nearly constant and only a small current is induced. Therefore there are just a few harmonics.

To get a better idea of the influence of the particle position relative to the FFP or the selection field, there is an overview illustrated in Figure 2.9, where three different positions for the nanoparticle sample are chosen. As can be seen, the particle response is highly dependent on the selection field applied. Therefore each particle has its own characteristic magnetic response. Supposing two different points are equidistant from the zero point x_0, the derivatives of the particle magnetisations have the same amplitude. As mentioned in [3], one way to distinguish between the two point samples is that the phases are rotated by 180 degrees due to the gradient. This means, even in this case, that a characteristic spectrum of magnetisation can be achieved.

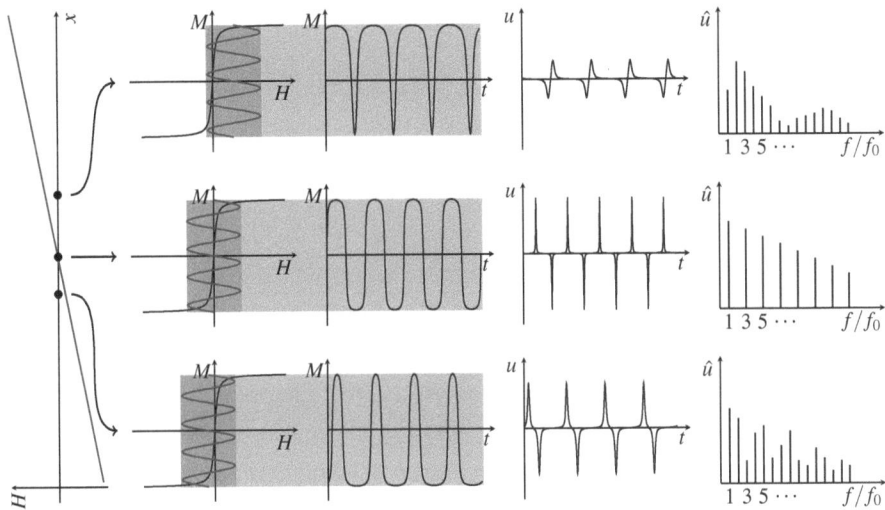

- Position of the sample

Figure 2.9: Visualisation of the influence of the gradient on the particle. The resulting signal changes relative to the different positions of the FFP.

2.2.4 Reconstruction

Image reconstruction in MPI can be understood as an inverse problem, whereby the distribution of the SPIO nanoparticles in a certain ROI is determined for a measured signal. Today, the approaches for reconstructing images in MPI can be separated into two main techniques: x-space reconstruction [18, 19, 20] and frequency-space reconstruction [15, 16, 39, 60, 73]. Both methods use the induced voltage $u(t)$ of the SPIO nanoparticles, as calculated in equation (2.40), to obtain a characteristic signal for reconstruction. While x-space reconstruction uses the induced voltage directly to reconstruct the image, the frequency space-based approach makes use of the Fourier transformation of the induced voltage $\hat{u}(t)$. Equation (2.40) can be rearranged so that it can be used as a basis for both methods. As shown in [32] the magnetisation can be written as

$$\vec{M}(\vec{r},t) = c(\vec{r})\overline{m}(\vec{H}(\vec{r},t)), \qquad (2.44)$$

where $\overline{m}$ is the mean magnetic moment of the SPIO nanoparticles. It is needed to motivate a more realistic calculation, due to the fact that the size of the particles in a tracer varies. By using equation (2.44), the calculation of induced voltage given by equation

(2.40) can be reformulated to

$$u(t) = -\mu_0 \frac{d}{dt} \int_\Omega c(\vec{r}) \overline{m}(\vec{H}(\vec{r},t)) \vec{h}(\vec{r}) \, d^3\vec{r} \tag{2.45}$$

$$= -\mu_0 \int_\Omega c(\vec{r}) \overline{m}'(\vec{H}(\vec{r},t)) \vec{H}'_D(\vec{r},t) \vec{h}(\vec{r}) \, d^3\vec{r}. \tag{2.46}$$

Using the x-space approach to reconstruct an image, there are four assumptions to be made regarding the imaging system [20]. First, a strong magnetic field must be generated with a unique and sufficient small FFP. Second, the SPIO nanoparticles can be adiabatically aligned and saturated by an external magnetic field. Third, the filtered low-frequency components of the nanoparticle signals in the receive signal are recoverable. Fourth, the generated drive fields must be homogenous. Consideration these assumptions allows further reformulations to the physical model of $u(t)$. It leads to

$$u(t) = -\mu_0 \vec{H}'_D(t) \vec{h} \int_\Omega c(\vec{r}) \overline{m}'(\vec{H}(\vec{r},t)) \, d^3\vec{r} \tag{2.47}$$

$$= -\mu_0 \vec{H}'_D(t) \vec{h} [c * \overline{m}](\vec{r}_{FFP}), \tag{2.48}$$

whereby it is shown in [60] that the convolution $c * \overline{m}$ can be ignored and an adjusted concentration $\check{c}$ can be introduced. Therefore, the equation (2.48) can be written as

$$u(t) = -\mu_0 \vec{H}'_D(t) \vec{h} \check{c}(\vec{r}_{FFP}). \tag{2.49}$$

The calculation of the adjusted concentration of the SPIO nanoparticles follows

$$\check{c}(\vec{r}_{FFP}) = -\frac{u(t)}{-\mu_0 \vec{H}'_D(t) \vec{h}}. \tag{2.50}$$

This implies that the induced voltage $u(t)$ is normalised with the movement speed of the drive field. Therefore, the normalised induced voltage is a kind of absolute value of $u(t)$, which can directly written into the image after gridding to the actual position of the FFP. The resulting image is blurred, which can be directly explained by the mathematical definition of the x-space formulation. The measured signal is convoluted with the point spread function (PSF) of the imaging system. The PSF used in the one-dimensional formulation is the derivative of the Langevin function as seen in equation (2.39) and Figure 2.5. One possible approach to compensate such blurring in the image is deconvolution. As an exemple technique, the Wiener deconvolution can be used to improve the image quality [17, 20]. To make the x-space technique more interesting for medical imaging it is shown in [19] that the formulation can be extended to 2D and 3D. However, it should be noted that no real-time imaging x-space MPI has been published to date.

As mentioned above, the frequency-space approach differs from the x-space technique because it makes use of the Fourier transformation of the induced voltage. In contrast to x-space, the measured concentrations are not directly written into the image. To model the connection between the measurement signal and the particle distribution a characteristic system function

$$S(\vec{r},t) = \overline{m}'(\vec{H}(\vec{r},t))\vec{H}'_D(\vec{r},t)\vec{h}(\vec{r}) \tag{2.51}$$

is established. Such a system function can be measured by moving a point source in a ROI [15]. Here, the system function is filled with the harmonic spectrum of the point source that is placed in every desired position in the considered ROI. Another possible way to set up a system function is to build a model that is directly correlated to the particle characteristics and the properties of the scanner used [35, 39]. In [23] a new approach known as the hybrid system function was presented which combined both techniques to speed up the data acquisition. Using a system calibration unit as published in [24] is another possible solution for a faster measuring process. Here, instead of moving the delta sample, the FOV is moved to the desired position. The most accurate way of determining the system function, however, is to use the measurement-based technique. This technique is the best method for taking account of inconsistencies and inaccuracies relating to the imaging system. Given the fact that there is a direct correlation between the spatial resolution of the image and the size of the system function, the higher the resolution the bigger the system matrix and consequently the longer the reconstruction time. A detailed description of the analysis of the system function can be found for example in [32]. However, the general mathematical problem which has to be solved is already given by equation (2.46) in combination with (2.51) and can be expressed by

$$u(t) = \int_\Omega S(\vec{r},t)c(\vec{r})\, \mathrm{d}^3\vec{r}. \tag{2.52}$$

By discretising this formula it can be written as

$$\vec{u} = \tilde{S}\vec{c} \quad \Leftrightarrow \quad \vec{c} = \tilde{S}^{-1}\vec{u}. \tag{2.53}$$

It can be seen that the calculation of the concentration of the SPIO nanoparticles can be solved by inverting the system matrix. This can be done by a singular value decomposition (SVD) [68], the conjugate gradient method (CG method) [71] or by using iterative methods such as the Kaczmarz method [26], which is also known as the algebraic reconstruction technique (ART) in CT [6]. As shown by Knopp et al. in [34, 37] the Kaczmarz method is highly suitable for MPI. This is because of a high rate of convergence [9] thanks to the fact that the rows in the MPI system matrix are nearly orthogonal [32].

2.2.5 Coil Geometries

MPI is an imaging modality based on magnetic fields. Measuring the spatial and temporal distribution of magnetic nanoparticles, these fields can be divided into selection fields and drive fields. The selection fields generate the FFP, which enable the spatial encoding, while the drive fields are responsible for moving the FFP along a trajectory in space.

To generate a selection field, a Maxwell coil pair can be used as it is for example shown in Figure 2.10. This coil pair consists of two opposite coils, through which the same current I flows. In this case, the flow direction of the current is in opposite directions. Therefore, the magnetic field which generated by the two coils is cancelled out in a specific region between the two coils; this region is named the FFP. Instead of using a Maxwell coil pair to generate the FFP, two permanent magnets can also be used with the magnetic north poles facing each other.

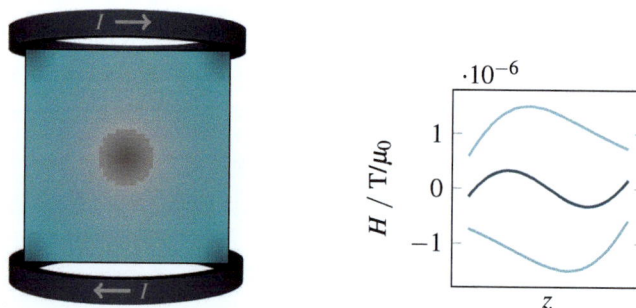

Figure 2.10: Visualisation of a Maxwell coil pair with the corresponding magnetic field. An FFP is generated in the middle of the two opposing coils.

As mentioned above, a drive field is used to move the FFP in space. Similar to the Maxwell coil configuration, an opposing pair of coils can be used, whereby unlike the

Maxwell coils, the current flows in the same direction. This coil configuration is known as a Helmholtz coil. An example visualisation of this coil pair is shown in Figure 2.11. Due to the fact that the current flows in the same direction, the magnetic fields of the coils are summated while overlapping. A higher magnetic field can therefore be generated. If the coils are at a distance to each other that is equal to their radius, the resulting magnetic field is homogenous in a certain region. In order to produce a three-dimensional movement in space, three Helmholtz coil pairs are needed which are mutually perpendicular. Alternatively, instead of using a Helmholtz coil, a solenoid coil can also be used.

 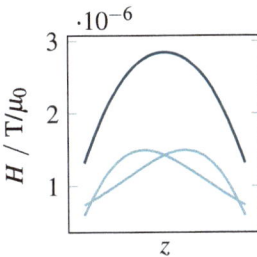

Figure 2.11: Example illustration of a Helmholtz coil pair. The magnetic fields of the coils overlap, which results in a higher magnetisation. For a certain distance between the coils, a homogenous magnetisation is generated.

Generally it is possible to distinguish between the generation and the detection of magnetic fields. Both can be performed by coils, which can be made of one piece or consist of a plurality of conductors using litz wire, which is used due to the skin effect [32]. In order to create a drive field, using both the one-piece coil and the litz wire coil is one possible solution to achieve a good result with respect to the magnetic field. For generation, coils made of litz wire are preferred. In contrast to the one-piece coils, the loss of power in the litz wire coil is lower and the achievable speed of movement of the FFP is substantially higher, which can be seen directly in relation to the temporal resolution. In addition to this, it is much easier to manufacture different coil geometries with sometimes difficult outer shapes by using litz wire. A more detailed examination of electrical conductors in magnetic fields and the resulting effects can be seen in [27]. However, it should be noted that this work focuses on the send coils and the receive coils are largely ignored.

Chapter 2 | Theoretical Principles

Figure 2.12: Visualisation of circular shaped coils. The appearance can be divided into an unchanged form (left), a curved one (middle) and a D-shaped coil (right).

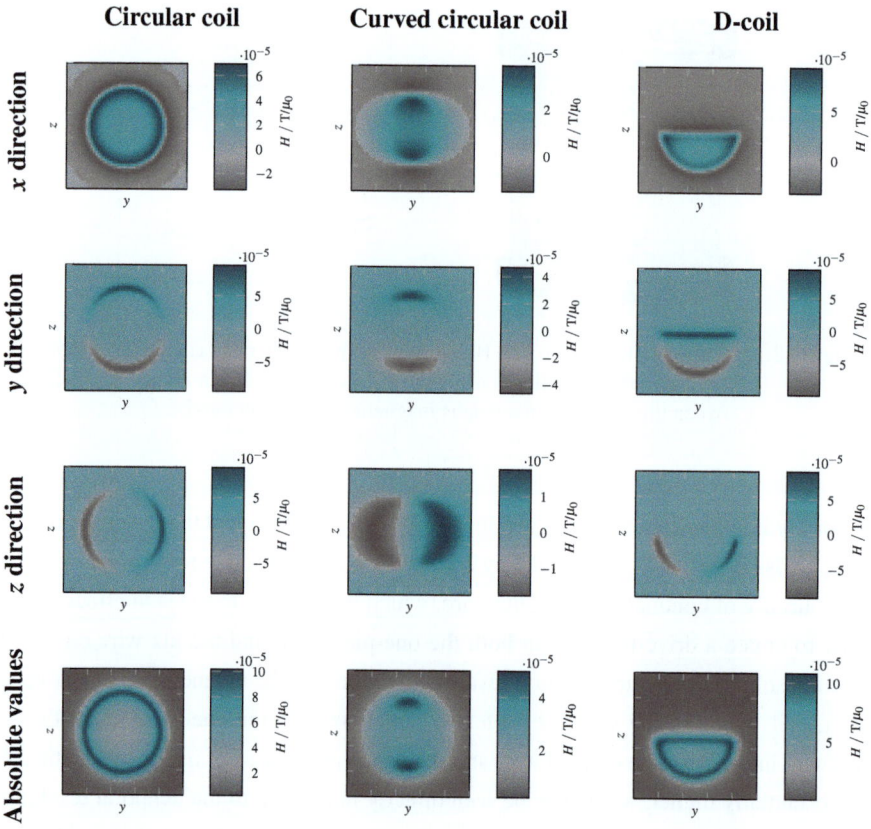

Figure 2.13: Magnetic fields of the circular based coils calculated on a 50 × 50 grid. Each column belongs to one of the geometries, while the field components for x, y and z direction and the absolute values are separated by the different rows.

Figure 2.14: Example illustration of rectangular shaped coils. A plane one is shown on the left side and a curved coil can be seen on the right side.

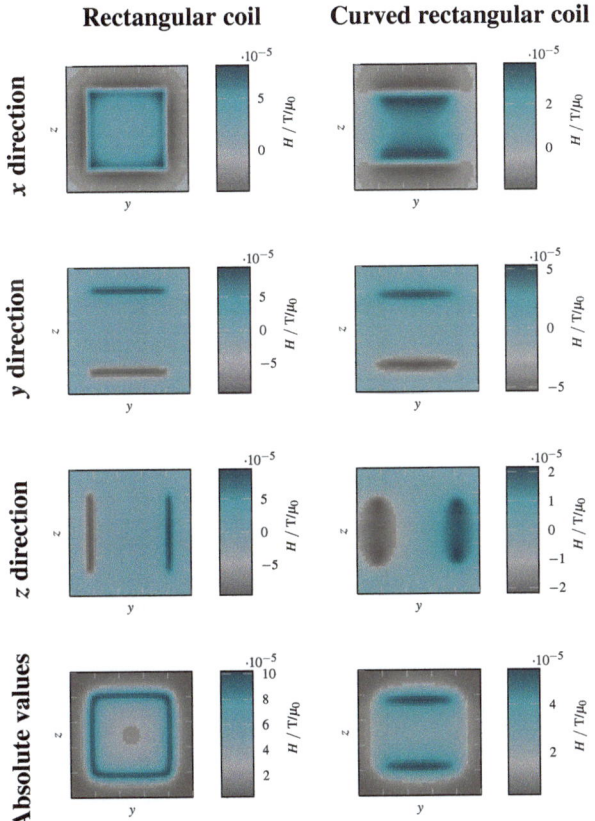

Figure 2.15: Magnetic fields of rectangular based coils calculated on a 50×50 grid. Each column belongs to one the geometries, while the field components for x, y and z direction and the absolute values are separated by the different rows.

Chapter 2 | Theoretical Principles

Different coil geometries can be used to generate a magnetic field. The chosen geometry of a coil depends greatly on the desired application and the technical possibilities. The simplest geometry consists of circular coils, as shown on the left side of Figure 2.12. Two variations of the circular configuration are illustrated in Figure 2.12 as well. In the middle there is a curved circular coil and on the right a coil shaped like a 'D'. The corresponding magnetic fields of each of the coil shapes mentioned are shown in Figure 2.13. The circular shaped coils have the advantage that they are easy to create using litz wire. Another advantage lies in the fact that the resulting magnetic field is homogeneous and rotationally symmetrical. The curved circular coils differ from the other two coils in that, due to the curvature, the distance to the object being viewed is smaller. The D-coils have neither rotational symmetry nor higher proximity to the object by curvature, but their geometry is advantageous. This advantage will be described later in the consideration of scanner topologies.

Another possible solution for the outer shape of a coil can be seen in Figure 2.14. Instead of using a circular geometry, the coils are rectangular shaped, with a differentiation being made between plane and curved rectangular shaped coils. A disadvantage of this coil shape is the fact that they are more difficult to achieve. Advantageous, however, is the fact that the performance needs to be less in order to achieve a similar result. The magnetic fields of these coil shapes are shown in Figure 2.15.

It must be noted that, depending on the shape of the coil, another magnetic field is generated, and thus the selection of the coil shape is very important. Important for this decision are the choice of scanner topology and the desired field of application. An example illustration of different scanner topologies is shown in Figure 2.16. Here, the scanners are not ordered chronologically, but according to the access pathway to a patient.

On the left is the conventional scanner topology, invented by Phillips. This scanner is constructed as a three-dimensional mice scanner [61, 75]. It consists of two opposing pairs of coils for the FFP movement in x and y direction, and a solenoid coil for the FFP movement in z direction. In this case the measuring area is located in the centre of the symmetric coil arrangement. Additionally above or below the coil in x direction, a permanent magnet is positioned in order to achieve an enlargement of the gradient. One advantage of such a scanner topology is that a homogenous magnetic field is generated. As a disadvantage may be considered the fact that it is a closed configuration where the access to the object or patient is very limited.

Figure 2.16: Presentation of the existing MPI scanner geometries. The conventional geometry for a mice scanner by Phillips is shown on the left; in the middle there is an open MPI scanner and on the right a single-sided device. In all of these geometries a visualisation of the receive coils is omitted.

The second scanner topology shown in Figure 2.16 is an open MPI scanner configuration similar to the one published in [66]. In contrast to the previous topology, this scanner produces the visualisation in the second and third dimension with two pairs of D-coils. As described in [65], using D-coils is advantageous, because of a lower power loss and the possibility of a very flatly design. The coil pairs are rotated by 90 degrees to each other. Looking at a single pair of D-coils, as shown in Figure 2.17, the direction of the current flow is selected such that it is along the straight segments in the same direction. The resulting magnetic field is orthogonal to the straight sections. One advantage of this scanner topology is the possibility of an interventional usage. Looking at the magnetic fields of this scanner topology, a distinction must be made between usage as a conventional scanner and as two single-sided scanners (as described below). The former generates a magnetic field with a good field quality in the middle region, which decreases towards the sides. Using it as two single-sided scanners means a good magnetic field quality directly near the coils and a lesser quality in the middle region. However, it should be noted that to date there has still been no detailed feasibility study on this scanner topology.

The single-sided scanner [64, 65] is the third scanner topology visualised in Figure 2.16. Unlike the conventional and the open MPI scanner that are mainly designed to be a full-body human scanner or at least used for larger areas of the body, this configuration is aimed at local application in the form of a hand-held device. It could therefore be possible to use this scanner design in the same manner as an ultrasound device which, as well as the benefits of being able to be used locally, brings with it the advantage of optimum patient access. Although in this case the resulting FOV and the penetration depth are substantially smaller than in the other two topologies, this is not to be seen as a disadvantage since the medical field of application is different. However, the resulting magnetic field is strongly inhomogeneous and the field strength decreases with distance from the scanner.

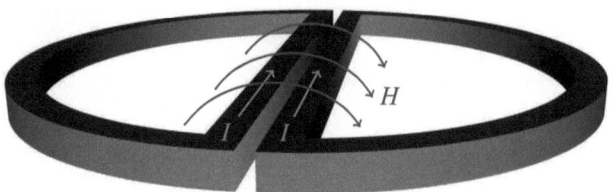

Figure 2.17: A pair of D coils can be used to generate a magnetic field H. The current flow of the coils needs to be contrary. With this configuration, a magnetic field is generated orthogonally to the straight sections.

3

Material and Methods

The following chapter is dedicated to the applied material and methods. The first part of this chapter contains information about a framework developed for simulating the physical processes in MPI (section 3.1). An extension of the simulation framework by including a new coil geometry is described in section 3.2. The chapter then elaborates on how the new coil geometry can be built (section 3.3). Finally, the conditions of the measurement process are discussed in section 3.4.

3.1 Simulation Framework

A framework created and implemented at the Institute of Medical Engineering at the University of Lübeck [32] can be used to simulate the physical processes in MPI. By using C++ the simulation framework combines a fast speed of execution with object oriented principles to easily extend the framework. It enables the user to arrange any number of different coil geometries and allows the integration of any kind of distribution of SPIO nanoparticles, the calculation of the magnetic field strength, the magnetisation of the nanoparticles and the induced voltage in the receive coils. Single coils and coil arrangements can be saved as XML files. Each of the files contains a defined description of the coil geometry, the position, the orientation and the flow direction of the current. The

Chapter 3 | Material and Methods

HDF5 format is used to save the calculated magnetic fields. This hierarchical data format enables efficient and flexible data storage. Furthermore, it is possible to use the files in most of the current programming languages. By combining the potential of the simulation framework with the advantages of scripting languages such as Matlab or Python, it is possible to carry out complex and automated calculations within an adequate timeframe. The framework has been implemented in such a way that different applications using the respective scripting language can be invoked and hence can be integrated into it.

When adding a new coil geometry to the framework, a distinction must be made between the visualisation of the coils and the calculation of the physical processes. As illustrated in Figure 3.1, both parts must be implemented separately. This means, as well as the correct visualisation of a new coil, the corresponding calculations of the above-mentioned physical processes must be performed as an individual operation.

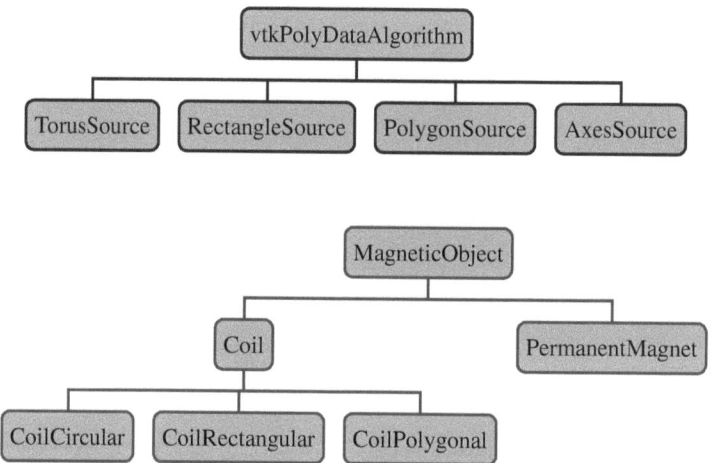

Figure 3.1: Part of the hierarchical structure of the simulation framework. The visualisation (green) and the calculations (red) are processed separately. For example, a circular shaped coil is visualised using the TorusSource method, while the calculations are performed with the CoilCircular method.

The visualisation is carried out by using the Visualization Toolkit (VTK), which is an open-source software system that consists of a C++ library [67]. VTK has a wide range of visualisation algorithms including three-dimensional image processing and visualisation. In the context of the simulation framework, it is used to visualise different coil geometries. By combining the coils it enables the user to simulate different scanner topologies as for example described in section 2.2.5. The graphical structuring of a coil

34

starts with the correct positioning of the point coordinates with respect to the selected coil form. In the next step, polygon surfaces are generated between each of the points. These surfaces are the actual visualisation of a coil. A simple illustration of how point coordinates are positioned and the respective connection traces for the polygon surfaces is shown in Figure 3.2. In addition to these two steps, a normal vector, perpendicular to the corresponding surface, is calculated for each of the point coordinates. It should be noted that for one point only one normal vector can be allocated. This normal vector influences the visualisation significantly with respect to the correct shading of surfaces. To calculate the shading correctly, each point of the visualised coil must be set twice, because there is always an edge between two neighbouring polygon surfaces and each of the edges requires a normal vector at this point. If the visualisation is performed without the normal vectors, the calculation speed can be increased at the expense of a smoother appearance, because only half the number of points are needed.

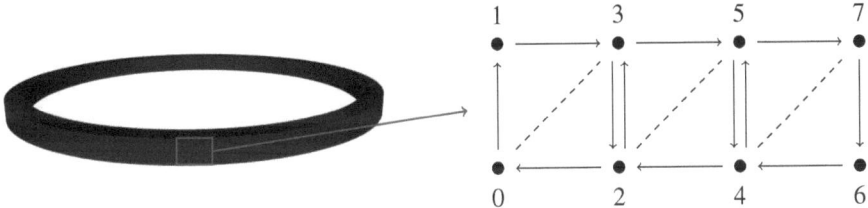

Figure 3.2: Sample positioning of a part of the point coordinates and the connection traces to create polygon surfaces for a circular coil. While the solid traces illustrate the connections between two point coordinates, the dashed lines represent separations of the polygon surfaces made by VTK to improve the performance.

The main underlying assumptions of the various calculation steps which are integrated in the framework are already presented in Chapter 2. The calculation of the magnetic field strength $\vec{H}$ is based on the Biot-Savart's law as introduced by equation (2.20). To analyse the field-generating properties of an electromagnetic coil, the sensitivity given by equation (2.41) can be used. As mentioned in [32], the magnetic field strength therefore can be calculated as

$$\vec{H}(\vec{r},t) = I(t)\vec{p}(\vec{r}). \tag{3.1}$$

This implementation of the magnetic field strength $\vec{H}$ offers the advantage of characterising the field-generating properties of an electromagnetic coil by considering the geometry alone. Additionally, a time-efficient calculation is provided by separating the temporal and spatial positions. Furthermore, this formulation for the magnetic field strength can be used to achieve an equally easy and advantageous calculation for the magnetisation of the SPIO nanoparticles and the induced voltage. A detailed analysis of the individual steps, from setting up a continuous model to the discretisation of the equations, can be found in [32].

3.2 Extension of the Simulation Framework

The following section provides information about a modification to the simulation framework and, based on this, an implementation of a new coil geometry.

3.2.1 Circle Segments

So far, the simulation framework provides a method for the visualisation of circular shaped coils. It is possible to choose between a full circle and different kinds of D-shaped coils. However, it should be noted that the actual visualisation is implemented only in this one function. This means, in order to realise further coils containing circular shaped components, it would be necessary to repeat the entire implementation work. Furthermore, the size of the desired circle segment can not be chosen arbitrarily and it is only capable of generating D-coils.

As an alternative implementation approach to this, the calculation of point coordinates and normal vectors is transferred into a separate function, being independent to the rest of the hierarchical visualisation structure. This externalisation offers a more flexible usage of part of circular shaped coils in other coil implementations. By implementing an option to choose a start angle and an end angle, it is possible to realise any kind of circular coil segment (see Figure 3.3).

Figure 3.3: Illustration of some possible segment sizes of a circular shaped coil.

To integrate such a circular segment into a coil and visualise the whole coil correctly, a distinction must be made between a vertical and a horizontal continuation at the end of the circular segment, as shown in Figure 3.4. Otherwise, the following segment of the coil would not have the appropriate width. If the next part of the coil is, for example, a straight part vertical to the current direction, as in D-shaped coils, the end of the circular segment has to be bevelled. If the next part is horizontal to the circular part, the end of the circular segment should be straight.

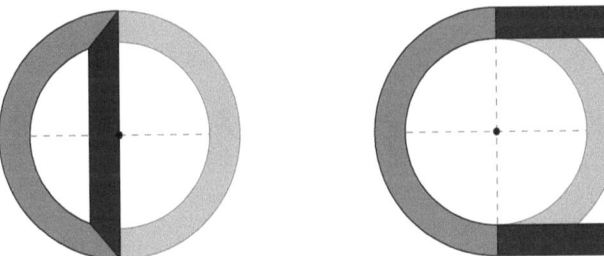

Figure 3.4: Representation of the two possibilities at the end piece of a circular segment. With a vertical continuation, a bevelled cut is needed (left side) and with a horizontal continuation the end is straight (right side).

The size of the angle of the bevelled cut must be calculated individually depending on the start and end angle combination of the circle segment. Calculating just a fixed angle of intersection only for one possible combination, representative of all the others, leads to a divergence in the visualisation. A sample illustration of the geometric conditions can be found in Figure 3.5. There are two angles to be calculated, one for the outer radius and one for the inner radius.

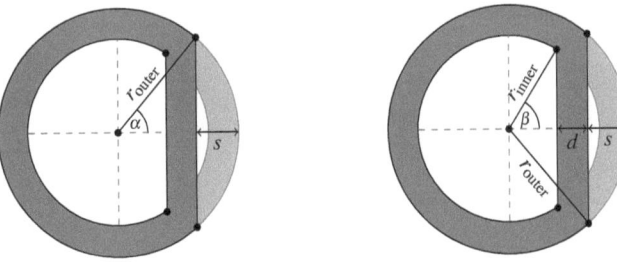

Figure 3.5: The figure on the left shows the geometric conditions with respect to the outer segment angle. The right side visualises them for the inner segment angle.

As shown in Figure 3.5, the calculation of the outer segment angle α is based on the outer radius r_{outer} and the segment height s. The segment height describes the distance between the outer edge of the coil and the outer straight line of the D-shaped coil. The calculation of the outer segment angle is given by

$$\alpha = \arccos\left(\frac{r_{outer} - s}{r_{outer}}\right). \tag{3.2}$$

To calculate the inner segment angle β, the inner radius r_{inner} and the coil thickness $d = r_{outer} - r_{inner}$ also have to be taken into account. Therefore, the calculation can be written as

$$\beta = \arccos\left(\frac{r_{outer} - s - d}{r_{inner}}\right). \tag{3.3}$$

With this modification of the simulation framework, it is much easier to integrate new coil geometries with circular parts.

3.2.2 Approximated Elliptical Coils

As described in section 3.1, the implementation of a coil geometry into the simulation framework is divided into a visualisation part and the calculation of the physical processes. To integrate a new coil geometry, both parts must be processed individually. The procedure is the same for the following description of the new coil geometry introduced here.

The new coil geometry is a combination of two circular shaped coils and a rectangular shaped coil. To be more precise, the coil consists of two semicircles connected by two straight parts in the middle, as shown in Figure 3.6. This new coil geometry is constructed to be an approximation of the geometric shape of an elliptical coil and is therefore referred to hereinafter as an approximated elliptical coil. In contrast to an elliptical coil, the choice of this coil form is based on a better feasibility in terms of geometric conditions and the structure of the simulation framework. At the same time it promises a good approximation of the elliptical shape.

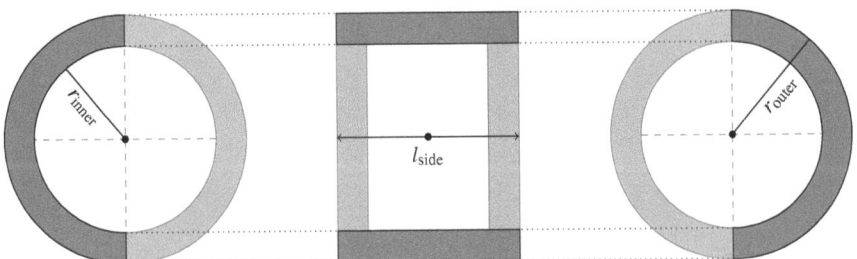

Figure 3.6: Visualisation of the individual components of an approximated elliptical coil. The coil consists of two semicircles connected by two straight parts in the middle.

As shown in Figure 3.7, an overlay of the approximated elliptical coil (green) and the elliptical coil (light grey) shows a significant intersection area. Therefore, the approximated elliptical coil seems to be a good approximation of the elliptical one. However, it should kept in mind that this coincidence is strongly dependent on the relationship between the two radii of the elliptical coil. An additional aspect is the side length of the approximated elliptical coil. The closer the outer shape is to a circle, the better the coincidence.

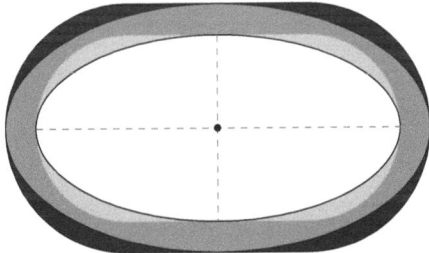

Figure 3.7: Overlay of a schematic illustration of an approximated elliptical coil (green) with an elliptical coil (light grey).

It is conceivable that the approximated elliptical coil can be used for example in an open MPI or single-sided MPI scanner topology. Therefore, it is necessary to implement D-shaped approximated elliptical coils to realise two-dimensional and three-dimensional imaging. In contrast to the rotationally symmetrical circular coils, where only the length of the segment height s has to be considered, the outer form of a D-shaped approximated elliptical coil also depends on the direction of the segment height. A distinction must be made between the segment height s_{long} for a D-coil along the long side of the coil and the segment height s_{short} for a D-coil along the short side of the coil. Both cases can be explained by Figure 3.8.

Chapter 3 | Material and Methods

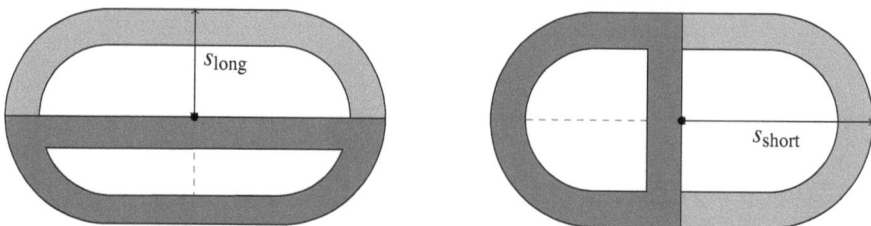

Figure 3.8: Possibilities for the realisation of a D-shaped coil based on an approximated elliptical coil.

An sample three-dimensional representation of all three variants of the approximated elliptical coil, as it can be visualised using the newly integrated method of the simulation framework, is shown in Figure 3.9.

Figure 3.9: Three-dimensional illustration of the different visualisation possibilities of an approximated elliptical coil. There is an unchanged form on the left side, a D-coil along the long side in the middle and a D-coil along the short side on the right.

With regards to the magnetic field strength $\vec{H}$ of an approximated elliptical coil, the calculation combines existing methods and the approaches introduced by the circle segments in section 3.2.1. The basic calculations of the magnetic field strength, as already mentioned in section 3.1, are based on the law of Biot-Savart as introduced by equation (2.20), the Maxwell equations given by equation (2.21) to (2.24) and the sensitivity of the coil given by equation (3.1). To calculate the magnetic field strength of the approximated elliptical coil, the coil is subdivided into four areas, whereby a distinction is made between straight sections and circular sections. The magnetic field strength of the circular sections is calculated with the conventional form of the above-mentioned equations. With regard to the calculation of the straight parts, it is possible to make several assumptions as described for example in [45, 46] to simplify it.

3.3 Realisation of the Simulated Coil Geometry

To validate the simulated approximated elliptical coil, the coil geometry is realised to perform comparative measurements of the magnetic field strength. The coil is made of rectangular shaped litz wire consisting of 2000 thin wire strands with a diameter of 0.05 mm with outer dimensions of 2.70 mm × 2.71 mm.

3.3.1 Geometry

The dimensions of the coil are an important factor in the motivation behind this work. Although it is rather negligible for initial comparisons, the dimensions of the constructed coil loan themselves to a problem of medical interest. The sample given situation is a single-sided MPI device using circular coils with a fixed size integrated into a patient table. As a benchmark, a circular coil of a single-sided device realised at the Institute of Medical Engineering at the University of Lübeck can be used (see Figure 3.10, grey coil). The use of approximated elliptical coils with regard to the geometry can be justified here by two approaches. First, the radius can be reduced by keeping a fixed length (see the red coil in Figure 3.10). If the circular coil used previously has been too wide for optimal patient access, this can be compensated for with this approach, while maintaining the length and therefore a similar imaging area. Second, the radius can be kept as fixed size, while extending the length of the sides, as shown by the green coil in Figure 3.10. This would make it possible to extend the imaging area by enabling patient access that is as good as using the circular coil. Both approaches have certain advantages, but due to the fact that the focus in this first study is on the comparative measurements of the magnetic field strength of simulated and real coils, the smaller red coil is realised to reduce the material to a minimum.

Chapter 3 | Material and Methods

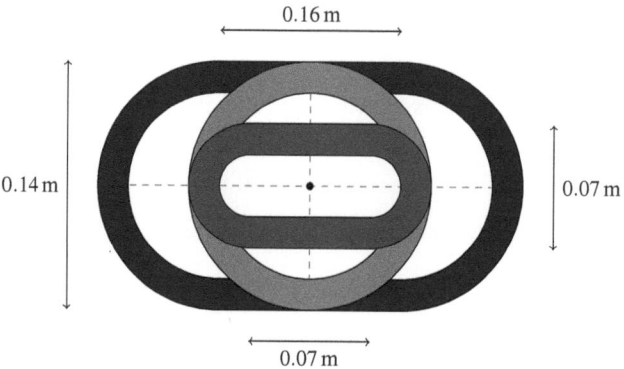

Figure 3.10: Comparison between two approaches to realise an approximated elliptical coil based on the geometric shape of a circular coil.

3.3.2 Construction of the Coil Mould

As a result of the fact that the realised coil is based on a new geometry and specific parameters, as described in section 3.3.1, a precisely suitable coil mould is built for these conditions. The coil mould is designed using the software 'SolidWorks', a three-dimensional computer-aided design (CAD) program. As shown in Figure 3.11, the mould consists of four different elements: bottom plate, mould, core and cover plate. The material used is polyoxymethylene (POM), which is characterised by good mechanical properties, high dimensional stability as well as excellent wear and friction behaviour [28]. In addition to the above-mentioned elements, the coil itself, realised using litz wire, is illustrated in Figure 3.11. To ensure permanent form stability, the litz wire coil is stuck together with two-component adhesive and is compacted under several tons of pressure.

3.4 Measurement Conditions

The measurements are performed with an isel Microstep Controller C 142-4.1 in combination with a Lake Shore Model 460 Gaussmeter, i.e. a 3-axis Hall effect sensor. The Hall effect can be described as an induced voltage in a current-carrying conductor proportional to the product of the magnetic field strength and the current [14, 25]. In contrast to a sensor consisting of coils and magnets, the Hall effect sensor is able to detect a signal in a static magnetic field. The measurement accuracy is $\pm 0.1\%$. The portions of

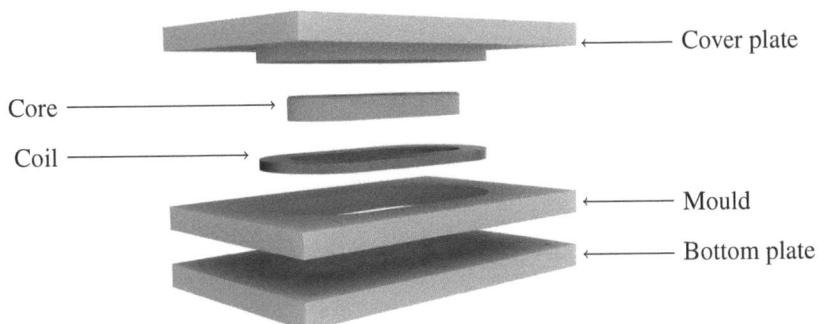

Figure 3.11: Three-dimensional visualisation of the constructed coil mould to build approximated elliptical coils using litz wire.

the sensor, where the sensitivity in x, y and z direction with regards to the magnetic field strength are the highest, are called the active area. The areas have a width of 0.5 mm and a length of 1 mm. The spatial arrangement of these areas can be seen in Figure 3.12. All three areas are perpendicular to each other within $\pm 0.5°$. It should be noted that the active z area is located 1.8 mm from the front side of the sensor, while the x area and y area are each positioned 2.08 mm away from the centre. The adhered tolerance is ± 0.25 mm. This issue must be factored into the later evaluation of the measured data, otherwise deviations in accuracy occur compared to the simulated data.

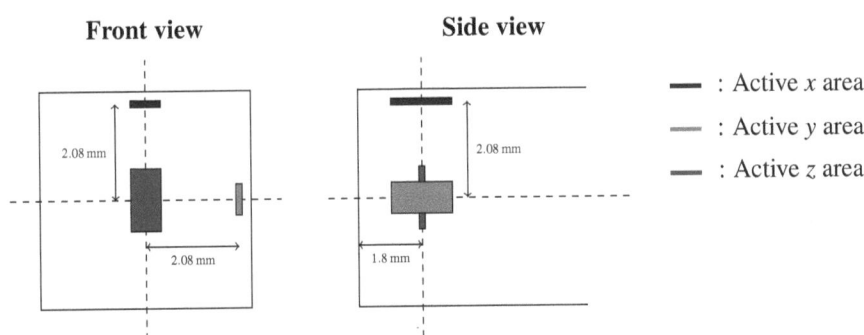

Figure 3.12: Graphical illustration of the spatial arrangement of the active areas inside the Hall effect sensor. A front view is shown on the left, while the side view can be seen on the right.

A simple experiment is carried out taking account of the later use of the measured data. A small two-dimensional area is scanned, with a permanent magnet being set in one of the corners (see left side of Figure 3.13). The aim is to obtain information about the

Chapter 3 | Material and Methods

relationship between the coordinate system of the isel system and the Hall effect sensor with respect to the positioning of the measured values in the image space. Comparing the position of the permanent magnet on the left side with the position in image space on the right side of Figure 3.13, it is apparent that the values are written at the correct positions in the image space.

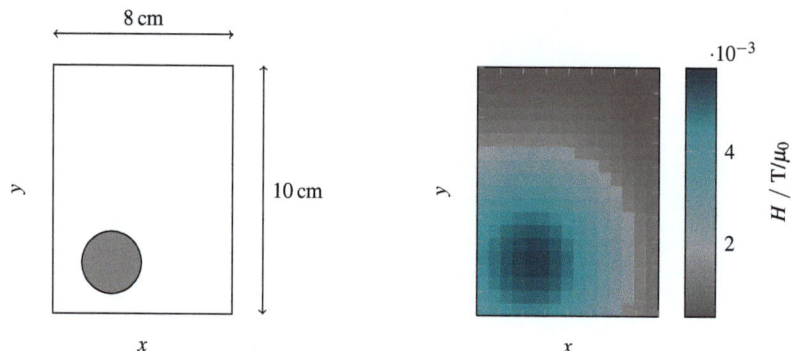

Figure 3.13: Schematic image of a defined area with a permanent magnet (grey) in the lower left corner (left side) and the corresponding results of the measured magnetic field (right side).

The whole measurement process, including the detection of the signals, filtering, conversions between analogue and digital signals and other electronic processing steps, has an inherent noise. This means that two independent measurements cannot have the same result. In order to approximately investigate the extent of the noise and how it could affect the measured data, several air measurements without generated magnetic fields are performed. It should be noted that environmental conditions and the magnetic field of the earth can affect the measurements. These influences are minimised by using a shielded copper cabin.

With respect to the resulting magnetic fields from the air measurements, the considerations are limited to the absolute values. In the images on the left and in the middle of Figure 3.14, the measured data of the noise affected air measurements can be comprehended. In addition to this, a difference image obtained by subtracting both images is shown on the right. The noise can be clearly detected in all three images.

To make further assumptions regarding the behaviour and the influence of the measured noise data, especially with the focus on a comparison between simulation results and the measured data in section 4.3, the frequency distribution of the difference image is

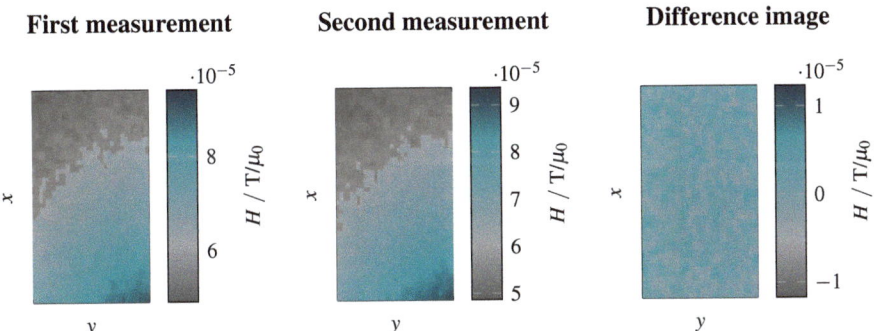

Figure 3.14: The images on the left and in the middle show the absolute values of two air measurements. Subtracting both images from each other results in a difference image shown on the right.

calculated. The width of the subdivided parts, also known as bins, is calculated according to [13, 69] by

$$b_w = 2 \cdot \frac{\text{IQR}(x_{\text{noise}})}{N^{1/3}}, \tag{3.4}$$

where IQR is the interquartile range (see [22]), x_{noise} are the values of the difference image and N is the number of considered values. By taking equation (3.4) into account, the number of bins is can be calculated by

$$b_n = \left\lceil \frac{\max(x_{\text{noise}}) - \min(x_{\text{noise}})}{b_w} \right\rceil. \tag{3.5}$$

The resulting frequency distribution is shown in Figure 3.15.

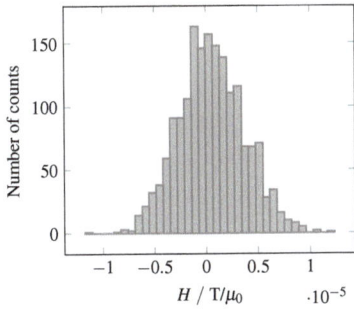

Figure 3.15: Frequency distribution of the difference of two air measurements.

It can be seen that the distribution of the values seems to be similar to the probability density function of a Gaussian distribution. This distribution can be calculated with

$$f(x) = \frac{1}{\sqrt{2\pi}} e^{-\frac{(x-\mu)^2}{\sigma^2}}, \qquad (3.6)$$

where μ is the mean value and σ is the standard deviation, which is usable as a measure of the measurement accuracy. This means that, in contrast to the simulation, the measurement can only be performed with an accuracy of approximately $\pm\sigma$. Fitting the Gaussian distribution to the measured data, as illustrated in Figure 3.16, shows that this approximation to the measured data is very accurate. For the standard deviation, a value of $\sigma = 4.427 \cdot 10^{-6}\, T\mu_0^{-1}$ is obtained.

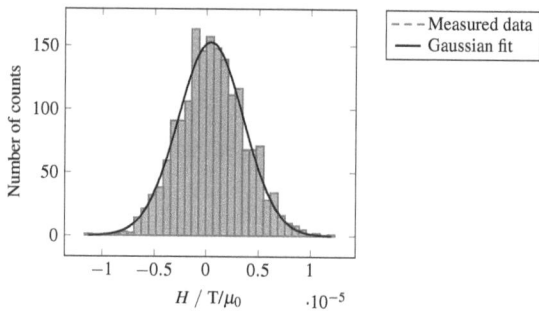

Figure 3.16: Representation of an analytical Gaussian fit to the frequency distribution of the difference of two air measurements.

3.5 Introduction of Error Metrics

To compare the difference between images, both subjective and objective criteria are used in practice. Metrics using subjective criteria are based on the human perception and are therefore not appropriate for determining differences between images. Objective metrics can initially be divided into three groups: reference [70, 72], reduced-reference [44, 47] and no-reference metrics [11, 53]. Reference-based methods consider information from the entire image while comparing the difference between two images. One of these images is often an image without any image errors and is therefore known as the reference image. For example, in CT such a reference image could be an acquired phantom, usable to validate reconstruction algorithms in terms of artefact reduction. For the application of reduced-reference metrics, only partial information is available from the reference image or used to evaluate the difference between images, for instance, to perform a real-time image comparison. However, if it is not possible to make use of a reference image, the no-reference methods can be used as an alternative approach. These metrics do not directly compare the difference in comparison to a reference image, but use selected image characteristics to perform a distinction independently of a reference image. An example is to compare the sharpness of different images, for instance, by considering the contrast.

Reference-based methods are used to consider the image differences between simulated and measured data. The methods used in this work are presented briefly below and serve as a comparative measure in chapter 4. These produce images of size $M \times N$ with $i = 1,\ldots,M$ and $j = 1,\ldots,N$, while $f(i,j)$ is the measured data and $g(i,j)$ represents the simulated data. To compare the differences between measured and simulated data correctly, the field components in x, y and z direction are considered separately.

One approach to comparing the difference between two images is the sum of squared differences (SSD), given by

$$\text{SSD}(f,g) = \sum_{i=1}^{M} \sum_{j=1}^{N} \sum_{x,y,z} (f(i,j) - g(i,j))^2. \tag{3.7}$$

The smaller the difference between the images, the smaller the value of the SSD.

Another approach, based on the idea of reference-based image comparison, is the normalised absolute differences (NAD), which can be calculated with

$$\text{NAD}(f,g) = \frac{\sum_{i=1}^{M} \sum_{j=1}^{N} \sum_{x,y,z} |f(i,j) - g(i,j)|}{\sum_{i=1}^{M} \sum_{j=1}^{N} \sum_{x,y,z} |f(i,j)|}. \tag{3.8}$$

The resulting values of the NAD are also proportional to the difference between the images.

Further, the calculation of the difference between two images can be described by the normalised root mean square error (NRMSE):

$$\text{NRMSE}(f,g) = \sqrt{\frac{\sum_{i=1}^{M} \sum_{j=1}^{N} \sum_{x,y,z} (f(i,j) - g(i,j))^2}{\sum_{i=1}^{M} \sum_{j=1}^{N} \sum_{x,y,z} (f(i,j) - \overline{f}(i,j))^2}}, \tag{3.9}$$

where

$$\overline{f}(i,j) = \frac{\sum_{i=1}^{M} \sum_{j=1}^{N} f(i,j)}{M \cdot N}. \tag{3.10}$$

The last reference-based metric presented here is the relative error (REL). With this, the difference between two images can be described by

$$\mathrm{REL}(f,g) = \frac{\sqrt{\sum_{i=1}^{M}\sum_{j=1}^{N}\Sigma_{x,y,z}(f(i,j)-g(i,j))^2}}{\sqrt{\sum_{i=1}^{M}\sum_{j=1}^{N}\Sigma_{x,y,z}f(i,j)^2}}. \tag{3.11}$$

The following also applies to the relative error: the higher the error, the bigger the difference between the images.

4

Experiments and Results

This chapter presents the experiments and results of the new coil geometry, the approximated elliptical coil as introduced in section 3.2.2. In the course of this, the magnetic fields and gradients of the approximated elliptical coils are compared to each other as well as to circular shaped coils. A comparison is also performed between the coils themselves and for different scanner topologies.

While section 4.1 deals with the presentation of the simulation of the magnetic fields generated by approximated elliptical coils, the measurement results of the implemented coils are shown in section 4.2. To validate the results of the simulations, the magnetic fields of section 4.1 and section 4.2 are compared to each other in section 4.3. A further validation of the simulated approximated elliptical coil is carried out in section 4.4 by comparing them to circular shaped coils. Additionally, this section evaluates the advantage of approximated elliptical coils compared to circular coils. Finally, in section 4.5 and section 4.6 a number of experiments are performed involving the extension of a single-sided scanner and an open MPI scanner, both described in section 2.2.5, with approximated elliptical coils. This section also provides the corresponding results with a focus on the gradient strength of the magnetic fields. For all of these comparisons, the error metrics introduced in section 3.5 are used as an objective criterion to compare the results.

Chapter 4 | Experiments and Results

4.1 Simulations

This section presents the simulation results of the approximated elliptical coils. The coil used for the simulations has an outer radius of 0.035 m, a side length of 0.07 m, a thickness of 0.0145 m and a length of 0.005 m. To simulate the two different types of D-shaped coils, the coil is modified by choosing the section height according to the outer radius or the side length. The simulations are at this point carried out with the same parameters as used for the measuring process, which is outlined in section 4.2. This is done in order to subsequently allow a better comparison between simulations and measurements. Performing the experiments with a direct current of 5 A, 10 A and 15 A enables the experiments to use a simple air cooling. The appropriate scale of the currents is chosen for a good comparison. The calculations of the simulated magnetic fields are performed using a plane rectangular shaped FOV, with a size of 0.14 m × 0.2 m discretised to a 35 × 50 grid. The coil is located in the xy plane with a distance of 0.016 m to the FOV.

The resulting magnetic fields generated by using a current of 5 A are shown in Figure 4.1, the ones at 10 A in Figure 4.2 and the ones at 15 A in Figure 4.3. All three figures have the same tabular structure: the approximated elliptical coil in column one and the approximated elliptical D-coils in column two and three. The different field components in x, y and z direction are visualised in the first three rows, while the last row shows the absolute values.

In order to compare the simulations to each other, the maximum and mean values for the different currents are calculated. The results can be obtained from Table 4.1. It can be seen that the maximum and mean values increase with an increasing current. Additionally, the values for the different field components and the absolute values are plotted in Figure 4.4. The course of the plotted maximum values increases linearly. By considering the mean values it can be seen that the values for the z direction and the absolute values increase, while the values for the x and y direction slightly decrease. The relationships between the maximum values in x, y and z direction as well as for the absolute values are concluded in Table 4.2. It is clarified that for all three field components as well as for the absolute values, the maximum field strength is approximately doubled between 5 A and 10 A and tripled between 5 A and 15 A.

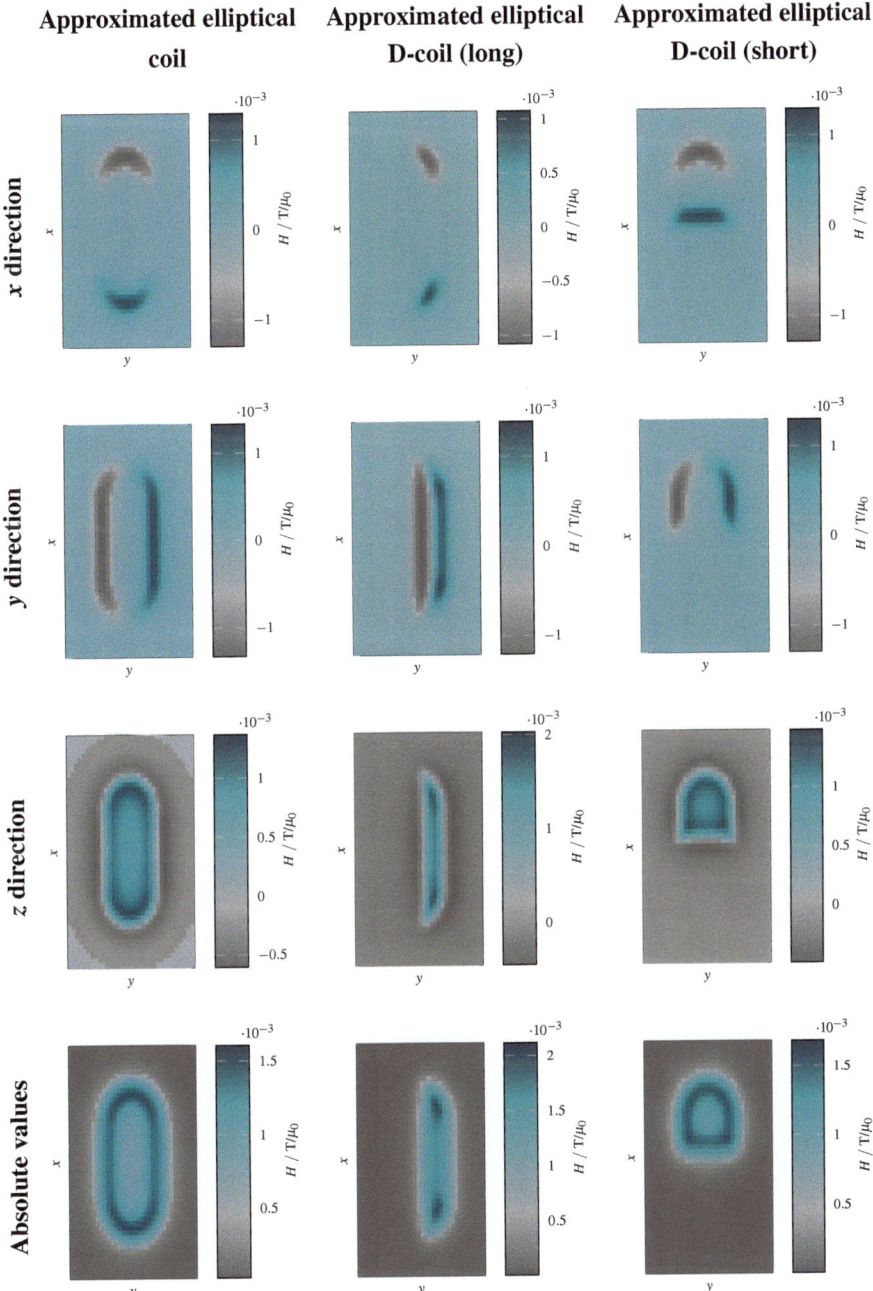

Figure 4.1: Magnetic fields of an approximated elliptical shaped coil and both D-coil variants, calculated on a 35×50 grid using a static current of 5 A. The following fields are presented: x direction (first row), y direction (second row), z direction (third row) and absolute values (fourth row).

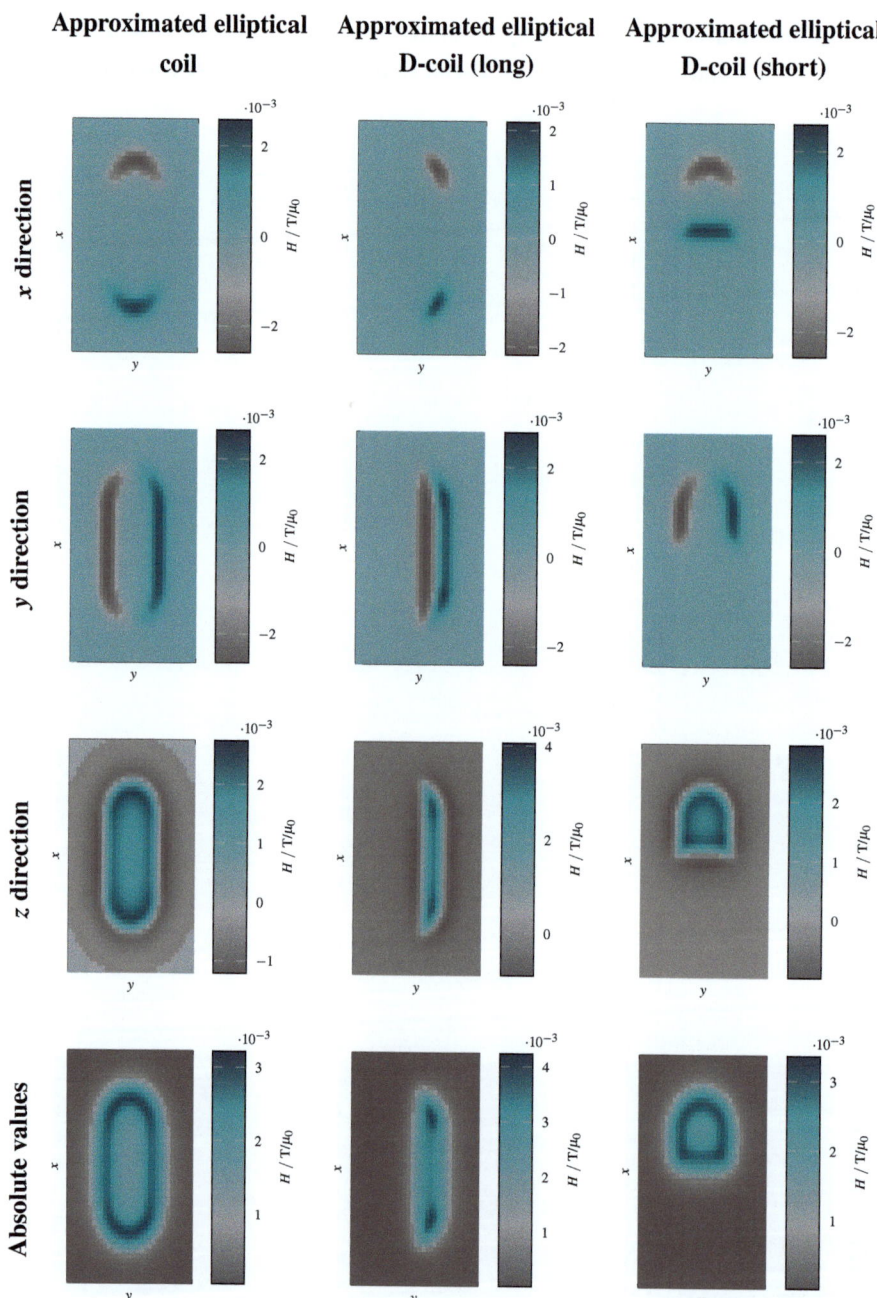

Figure 4.2: Magnetic fields of an approximated elliptical shaped coil and both D-coil variants, calculated on a 35×50 grid using a static current of $10\,\text{A}$. The following fields are presented: x direction (first row), y direction (second row), z direction (third row) and absolute values (fourth row).

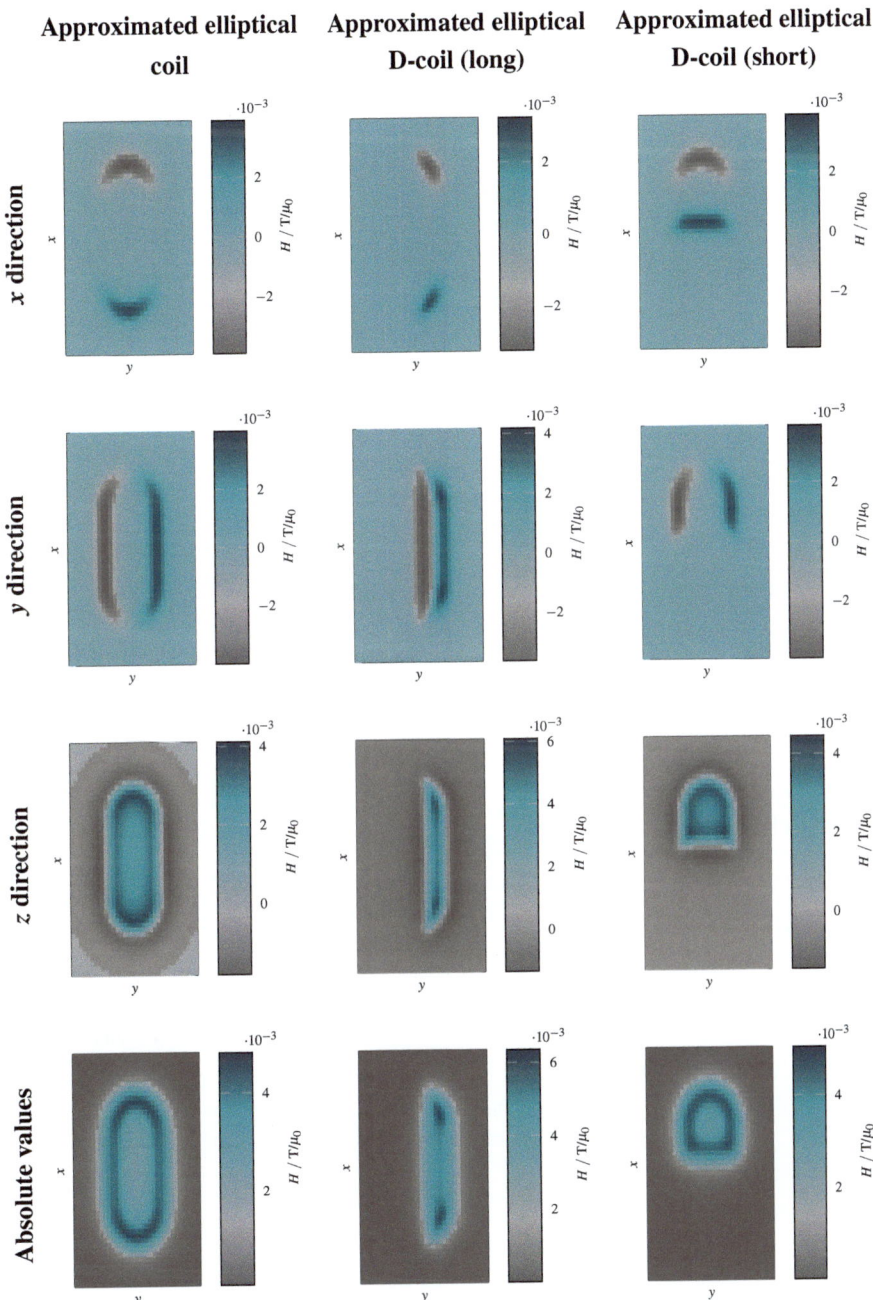

Figure 4.3: Magnetic fields of an approximated elliptical shaped coil and both D-coil variants, calculated on a 35×50 grid using a static current of 15 A. The following fields are presented: x direction (first row), y direction (second row), z direction (third row) and absolute values (fourth row).

Chapter 4 | Experiments and Results

Table 4.1: Maximum and mean magnetic field strength values of the simulated magnetic fields for the different field components and the absolute values using a current of 5 A, 10 A and 15 A.

		5 A / T/μ_0	10 A / T/μ_0	15 A / T/μ_0
Maximum	x direction	$4.9474 \cdot 10^{-4}$	$9.7720 \cdot 10^{-4}$	0.0015
	y direction	$5.3910 \cdot 10^{-4}$	0.0011	0.0016
	z direction	$6.6377 \cdot 10^{-4}$	0.0013	0.0020
	Absolute values	$7.0737 \cdot 10^{-4}$	0.0014	0.0021
Mean	x direction	$-4.5343 \cdot 10^{-8}$	$4.2238 \cdot 10^{-8}$	$1.5896 \cdot 10^{-7}$
	y direction	$-9.0344 \cdot 10^{-7}$	$-1.7617 \cdot 10^{-6}$	$-2.8901 \cdot 10^{-6}$
	z direction	$8.1216 \cdot 10^{-5}$	$1.6222 \cdot 10^{-4}$	$2.4339 \cdot 10^{-4}$
	Absolute values	$2.9218 \cdot 10^{-4}$	$5.8075 \cdot 10^{-4}$	$8.7162 \cdot 10^{-4}$

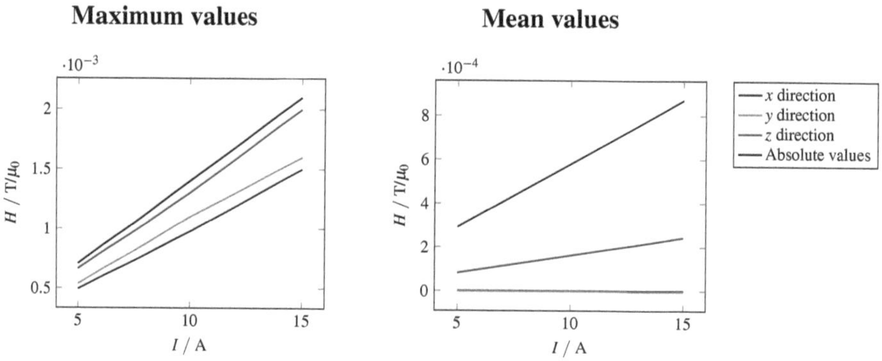

Figure 4.4: Visualisation of maximum and mean magnetic field strength of the simulated data for 5 A, 10 A and 15 A. The magnetic field strength is separated in its components: x direction, y direction, z direction and the absolute values.

Table 4.2: Relationships between the maximum values of the magnetic field components in x, y and z direction and the absolute values for 10 A / 5 A as well as 15 A / 5 A.

	10 A / 5 A	15 A / 5 A
x direction	1.9751	3.0319
y direction	2.0404	2.9679
z direction	1.9585	3.0131
Absolute values	1.9791	2.9687

4.2 Measurements

The measurements are performed by using a coil made of litz wire as described in section 3.3.1. As already mentioned in section 4.1, the built coil has the same dimensions as the simulated coil and the currents used are also the same. The resulting measured magnetic fields of the approximated elliptical coil, calculated for a FOV with a size of $0.14\,\text{m} \times 0.2\,\text{m}$ discretised to a 35×50 grid using a direct current of 5 A, 10 A and 15 A, are shown in Figure 4.5. The field components are shown in the different rows, the x direction in the first row, the y direction in the second row, the z direction in the third row and the absolute values in the fourth row.

A calculation of the maximum values as well as the mean values of the magnetic field components in x, y and z direction and the absolute values are illustrated in Table 4.3. With an increasing current the values of the maxima increase as well. Regarding the absolute mean values, the numbers also increase for the z direction and the absolute values, while decreasing slightly for the x and y direction. Each value is calculated for 5 A, 10 A and 15A. In order to feature a visual comparison of the maximum and mean values, the resulting values are plotted in Figure 4.6. The courses obtained from Table 4.3 can be seen here as well. To get an objective measure of the relationship between the different currents, the maximum values are divided by each other. The results are shown in Table 4.4. The values of the x and y direction as well as the absolute values are doubled or tripled between 5 A and 10 A or 15 A and 5 A. The values for the z direction are approximately tripled between 5 A and 10 A and the values for 15 A are six times as high as for 5 A.

Chapter 4 | Experiments and Results

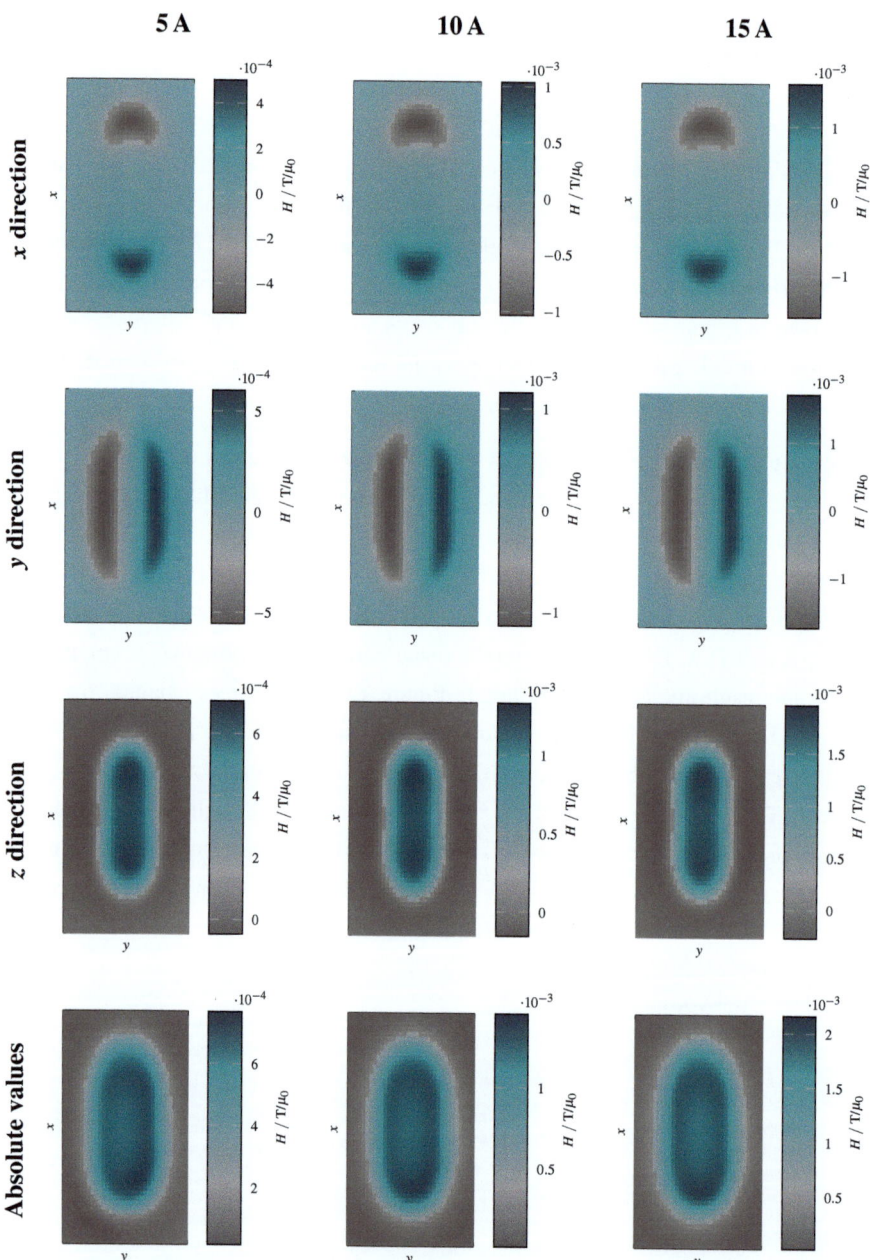

Figure 4.5: Resulting measured magnetic fields of an approximated elliptical shaped coil, calculated on a 35×50 grid using a direct current of 5 A, 10 A and 15 A. The field components are shown in the different rows: the x direction in the first row, the y direction in the second row, the z direction in the third row and the absolute values in the fourth row.

4.2 Measurements

Table 4.3: Maximum and mean magnetic field strength values of the measured magnetic fields for the different field components using 5 A, 10 A and 15 A.

		5 A / T/μ_0	10 A / T/μ_0	15 A / T/μ_0
Maximum	x direction	$5.2900 \cdot 10^{-4}$	0.0010	0.0016
	y direction	$5.5400 \cdot 10^{-4}$	0.0011	0.0017
	z direction	$4.7000 \cdot 10^{-5}$	$1.5500 \cdot 10^{-4}$	$2.7367 \cdot 10^{-4}$
	Absolute values	$7.6718 \cdot 10^{-4}$	0.0015	0.0022
Mean	x direction	$1.9217 \cdot 10^{-5}$	$1.0269 \cdot 10^{-5}$	$7.0682 \cdot 10^{-6}$
	y direction	$-2.4876 \cdot 10^{-5}$	$-9.5636 \cdot 10^{-6}$	$1.5058 \cdot 10^{-5}$
	z direction	$-1.5363 \cdot 10^{-4}$	$-2.2698 \cdot 10^{-4}$	$-2.9989 \cdot 10^{-4}$
	Absolute values	$3.0243 \cdot 10^{-4}$	$5.7224 \cdot 10^{-4}$	$8.6628 \cdot 10^{-4}$

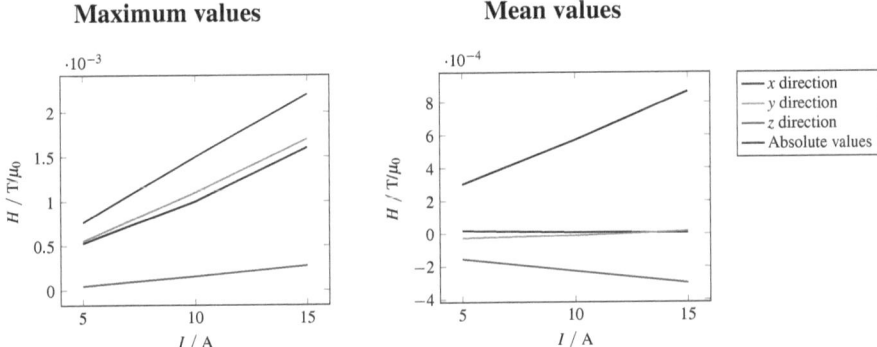

Figure 4.6: Visualisation of maximum and mean magnetic field strength of the measured data for 5 A, 10 A and 15 A. The magnetic field strength is separated in its components: x direction, y direction, z direction and the absolute values.

Table 4.4: Relationships between the maximum values of the magnetic field components in x, y and z direction and the absolute values for 10 A / 5 A as well as 15 A / 5 A.

	10 A / 5 A	15 A / 5 A
x direction	1.8904	3.0246
y direction	1.9856	3.0686
z direction	3.2979	5.8228
Absolute values	1.9552	2.8676

Chapter 4 | Experiments and Results

4.3 Comparison between Simulation and Measurement

In order to validate the simulated magnetic fields, a comparison with the measurements of the implemented coil is performed. To compare the simulated and measured magnetic fields correctly, the initial conditions must be adapted first. This is because the measurements of the approximated elliptical coil can differ from the simulations especially with respect to the positioning of the coil or the FOV. For example, if the Hall effect sensor is not centred directly in the middle of the coil or the coil is not positioned flat, but with a slightly slant angle to the surface of the measurement table, the magnetic fields are rigidly transformed. A simple approach to correct such a difference between a measured and a simulated magnetic field is to adapt the FOV of the simulations by a translation in x, y and z direction. Admittedly, a difference caused by a slant angle to the measurement table cannot be corrected by this method, but it serves as a good first approach for correcting the difference between the magnetic fields. To solve such a problem completely, there are several different techniques for image registration that can be used [54, 55]. Due to the complexity of this topic, this work is limited to the simple approach. The difference between the magnetic fields is minimised using the Matlab function `fminsearch` based on the simplex method published by Nelder and Mead [59]. In combination with this, the SSD introduced by equation (3.7) is used as a distance metric. The resulting recalculated centre positions of the FOV for the simulated magnetic fields for 5 A, 10 A and 15 A are shown in Table 4.5. Compared to the initial distance in z direction of 0.016 m, the position of the FOV is corrected by about 0.0003 m towards the coil surface. In y direction the centre position is moved approximately 0.0019 m. The misalignment in x direction varies, but is limited to a maximum shift of $-2.1272 \cdot 10^{-4}$ m.

Table 4.5: Resulting values of the centre position of the FOV for the fitting of simulation and measurement.

	5 A	10 A	15 A
x direction / m	$-1.8304 \cdot 10^{-4}$	$8.4730 \cdot 10^{-5}$	$-2.1272 \cdot 10^{-4}$
y direction / m	-0.0019	-0.0019	-0.0021
z direction / m	0.0157	0.0158	0.0158

To extend the comparison between simulation and measurement, the objective error metrics introduced in section 3.5 are applied for this case. The results of these metrics are concluded in Table 4.6. The error calculated by the NAD, the NRMSE and the REL are approximately in the same range of values and decrease by choosing a higher electrical current for simulation and measurement. The SSD values are much smaller than the other error metric values and increase slightly with a higher current. The corresponding magnetic fields of the adapted simulation and the measurement as well as the resulting images of the REL between simulation and measurement are illustrated in Figure 4.7.

Table 4.6: Comparison between simulated and measured magnetic fields of an approximated elliptical coil.

	5 A	10 A	15 A
SSD	$1.3550 \cdot 10^{-5}$	$1.4363 \cdot 10^{-5}$	$2.0815 \cdot 10^{-5}$
NAD	0.2828	0.1509	0.1263
NRMSE	0.2374	0.1239	0.0992
REL	0.2236	0.1191	0.0961

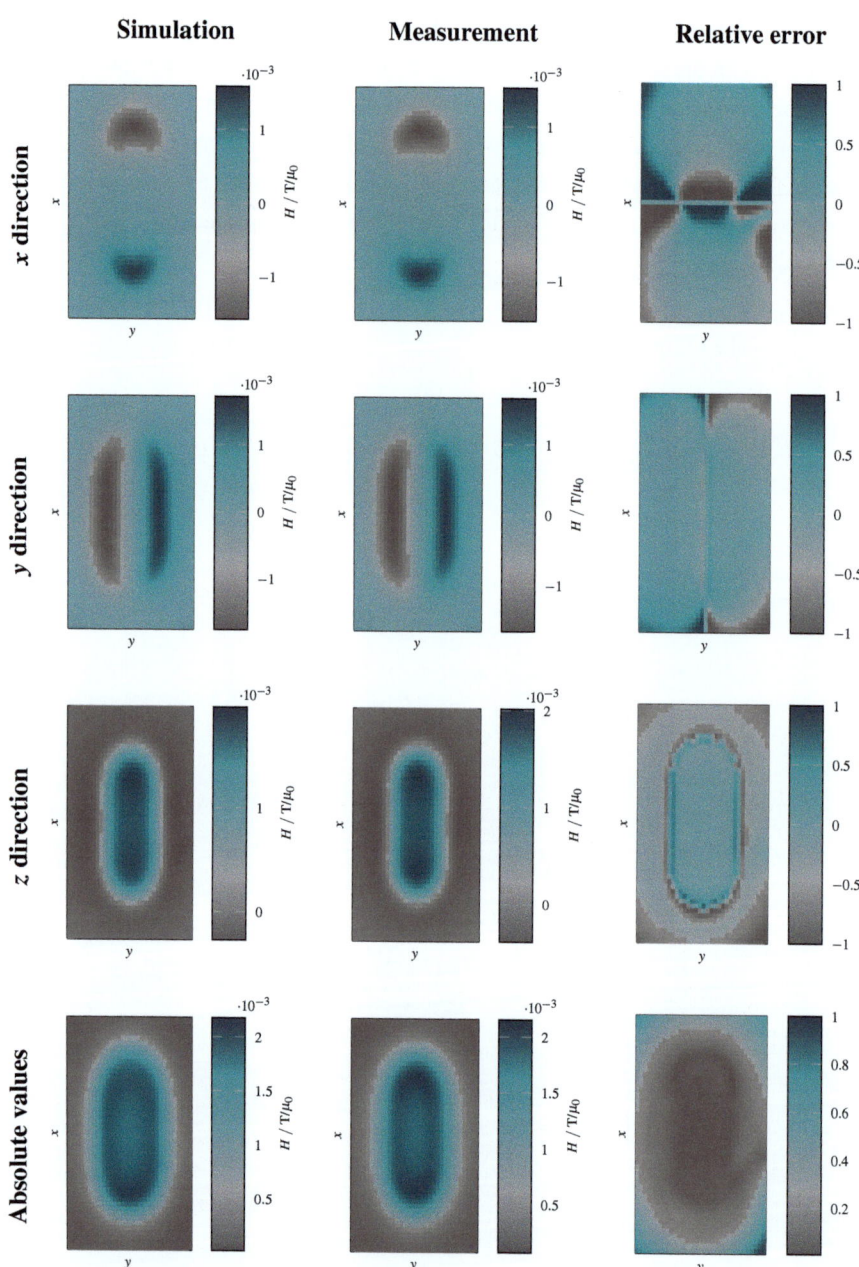

Figure 4.7: Visualisation of the simulated magnetic field components and the absolute values compared to the measured data. The simulated data is fitted to the measured data. The relative error is also shown.

4.4 Circular Coils versus Approximated Elliptical Coils

With a focus on a possible subsequent applicability, the approximated elliptical coils are compared with circular shaped coils. Regarding the different magnetic fields, the differences and the benefits of the coils are evaluated. In this context, a further validation of the approximated elliptical coil is performed in section 4.4.1 by using a side length of 0 m as well as an extension of this circular shaped coil by increasing the side length up to 1 m as presented in section 4.4.2.

The radius of the coils is 0.1 m, the thickness is 0.01 m and the length is set to 0.01 m. For this experiment the coils are positioned in the yz plane using a direct current of 5 A. The orientation of the FOV is varied between a parallel orientation with respect to the coil and a vertical orientation. In both cases the FOV does not intersect with a part of the coil. The horizontal FOV is 0.25 m × 0.25 m discretised to a 50 × 50 grid with a distance of 0.0157 m to the coil. The vertical FOV is 0.4 m × 0.15 m discretised to a 80 × 30 grid positioned in the centre of the coil.

4.4.1 Validation

To validate the simulation, the side length of the approximated elliptical coil is set to 0 m. The resulting circular shaped coil based on the approximated elliptical one is then compared to the existing circular coil by considering the differences between the magnetic fields. As an objective possibility for comparing the magnetic fields, the error metrics introduced in section 3.5 are used. The results of the comparison of the magnetic fields generated by the circular coils are presented in Table 4.7. The differences between the magnetic fields are small enough to be neglected, with the resulting values for the SSD being even below the machine accuracy that is $2.2204 \cdot 10^{-16}$ and can therefore be set to zero.

Table 4.7: Comparison between the magnetic fields of a circular shaped coil and an approximated elliptical coil with a side length of 0 m.

	Horizontal FOV	Vertical FOV
SSD	$4.6062 \cdot 10^{-19}$	$1.2353 \cdot 10^{-31}$
NAD	$1.2268 \cdot 10^{-7}$	$2.9649 \cdot 10^{-14}$
NRMSE	$3.4605 \cdot 10^{-7}$	$5.4486 \cdot 10^{-13}$
REL	$3.3006 \cdot 10^{-7}$	$3.6966 \cdot 10^{-13}$

4.4.2 Extension

The other experiment performed in this context is the variation of the side length of the approximated elliptical coil and the resulting effect on the magnetic fields with a focus on the maximum and mean values. The side length of the approximated elliptical coil is set in a range from 0 m to 1 m with a step width of 0.01 m. The results for the horizontal FOV are shown in Figure 4.8 and the results of the vertical FOV can be seen in Figure 4.9. Focusing on the horizontal FOV, the maximum values for a side length of 0 m of the approximated elliptical coil are almost the same as the circular ones. With an increasing side length, the maximum values decrease, while the course seems to near a limit. Considering the mean values, the same behaviour can be observed for the x direction and the absolute values. With respect to the y and z direction, the values are nearly constant and equal to the values of the circular coil because the difference is below the machine accuracy. For the vertical FOV, the maximum values in x and z direction and the absolute values as well as the course of the mean values in x direction and for the absolute values are similar to the ones of the horizontal FOV. Regarding the maximum and mean values in y direction and the mean values in z direction, the differences are below the machine accuracy and can therefore be neglected.

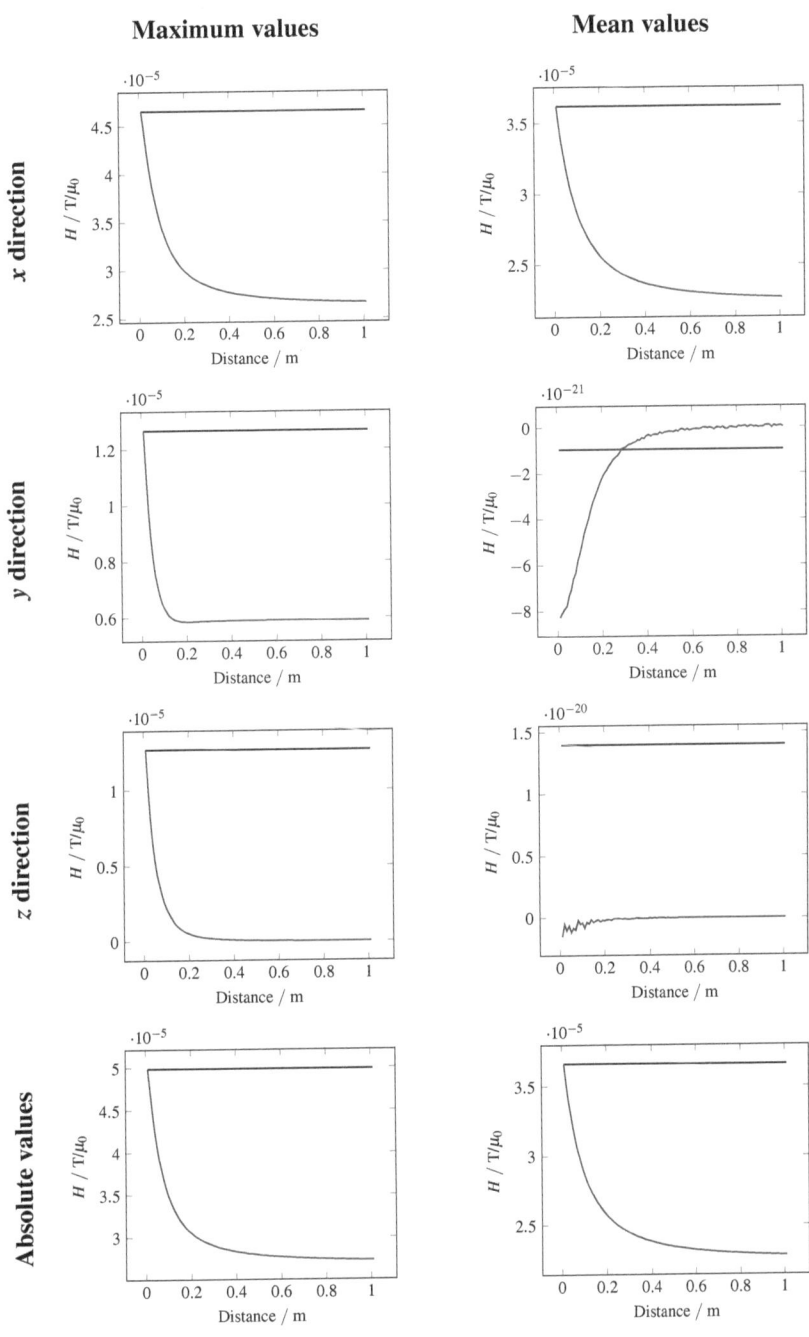

Figure 4.8: Graphical illustration of the difference between a circular shaped coil (green) and an approximated elliptical coil (red) with a side length from 0 m to 1 m gradually increased by 0.01 m. Considered are the maximum and mean values of the generated magnetic fields using a current of 5 A with a FOV parallel to the coils.

Chapter 4 | Experiments and Results

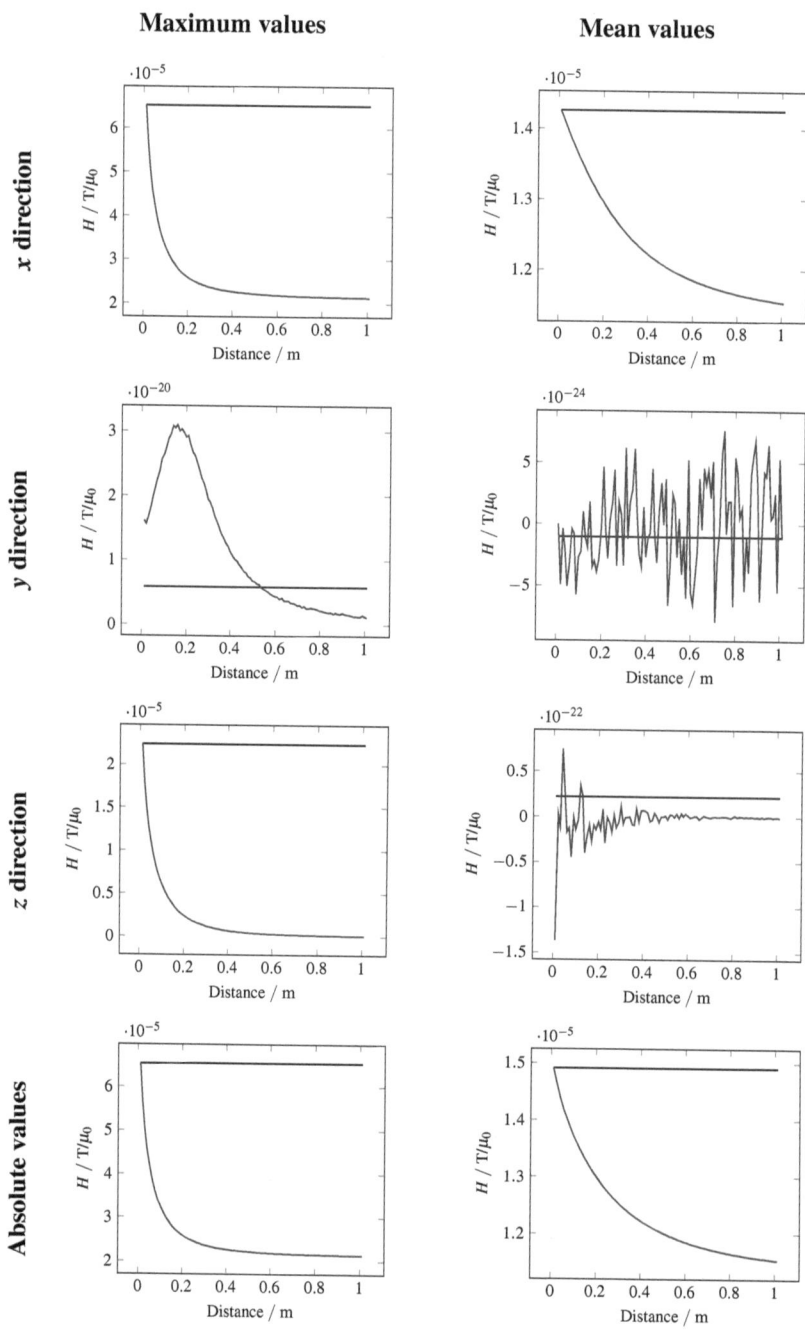

Figure 4.9: Graphical illustration of the difference between a circular shaped coil (green) and an approximated elliptical coil (red) with a side length from 0 m to 1 m gradually increased by 0.01 m. Considered are the maximum and mean values of the generated magnetic fields using a current of 5 A with a FOV vertical to the coils.

4.5 Application within a Circular Single-Sided Scanner

In this section the possible application of approximated elliptical coils for building a single-sided scanner is investigated. By setting the side length of the approximated elliptical single-sided scanner to 0 m, the usability of this coil geometry in such a scanner topology can be validated. The results of the validation are presented in section 4.5.1. To extend the basic validation, the side length of the approximated elliptical scanner is increased. The experimental results can be observed in section 4.5.2. The experiments performed here are focused on the course of the magnetic field strength and the gradient with increasing distance to the scanner. A sample illustration of a circular based single-sided scanner is shown on the left side of Figure 4.10, the single-sided scanner based on approximated elliptical coils being shown on the right side. The dimensions of both scanners are based on an already realised circular shaped scanner built at the Institute of Medical Engineering at the University of Lübeck [64]. It should be mentioned that the dimensions of the simulated scanners are not exactly the same dimensions as from the built scanner because of a different litz wire used. Due to the fact that for this comparison only a static magnetic field is generated, the description of the dimensions is limited to the green coils shown in Figure 4.10.

The radius of the outer coil is 0.07 m, the radius of the inner coil is 0.03 m, both coils have a thickness of 0.01625 m and a length of 0.0155 m. The scanners are located in the yz plane, while using a direct current of 55 A for the outer coil and -65 A for the inner coil. The FOV is centred to the coil with a distance of 0.011 m in x direction. The dimensions are 0.03 m $\times$ 0.03 m $\times$ 0.03 m calculated on a $15 \times 15 \times 15$ cubic shaped grid.

4.5.1 Validation

By setting the side length of the approximated elliptical single-sided scanner to 0 m, the difference between the generated magnetic fields to the circular shaped scanner can be observed. Using the error metrics introduced in section 3.5, the resulting differences are: SSD = 0.0029, NAD = 0.0695, NRMSE = 0.0749 and REL = 0.0721. The error values between the magnetic fields are all under 10 %, the SSD value is even under 1 %.

Chapter 4 | Experiments and Results

Figure 4.10: Comparative illustration of a conventional circular shaped single-sided scanner and a single-sided scanner realised by using approximated elliptical coils.

The results with respect to the magnetic field strength and the gradient strength for the comparison of two circular shaped single-sided scanners, with one being based on approximated elliptical coils, are shown in Figure 4.11. For both, the magnetic field strength in x direction decreases with an increasing distance to the scanner. In y and z direction, the magnetic field strength switches sign after crossing zero, while the course is nearly symmetrical. The highest values of the gradient in x direction are directly next to the scanner surface, while in y and z direction the highest gradient is at the position of the zero-crossing. The gradients in the FFP are shown in Table 4.8. Both the gradient and the strength of the magnetic field have very similar values. The distance of the FFP to the scanner is the same.

Table 4.8: Resulting values for the magnetic field strength and the gradient strength of a single-sided MPI scanner using 55 A for the outer coil and −65 A for the inner coil. One scanner is based on an existing circular coil, while the other scanner is also a circular shaped scanner based on approximated elliptical coils with a side length of 0.00 m.

		Distance / m	Gradient / T/m	H / T/μ_0
Circular coil	x direction	0.02	-0.6378	$-1.679 \cdot 10^{-5}$
	y direction	0.016	0.315	$-6.258 \cdot 10^{-19}$
	z direction	0.016	0.315	$1.546 \cdot 10^{-17}$
Approximated elliptical coil	x direction	0.02	-0.6483	$5.193 \cdot 10^{-4}$
	y direction	0.016	0.3206	$-1.178 \cdot 10^{-17}$
	z direction	0.016	0.3206	$-2.552 \cdot 10^{-18}$

4.5.2 Extension

To compare both scanner geometries and to evaluate if the approximated elliptical coils have an advantage over the circular shaped coils, the side length of the approximated elliptical scanner is varied between 0 m and 0.1 m for the outer coil and 0 m and 0.09 m for the inner coil.

As a sample scanner configuration, the side length is increased to 0.03 m and 0.02 m as shown in Figure 4.12. The course of the curves is similar to the ones shown in Figure 4.11. In comparison to the values of the circular single-sided scanner, the curves of the approximated elliptical scanner are flatter and not as high as the circular ones. The corresponding gradient in the FFP as well as the field strength values are shown in Table 4.9. It can be seen that the gradient decreases in contrast to the circular based scanner, while the field strength is nearly the same. The distance of the FFP in the x direction is increased compared to the circular single-sided scanner.

Table 4.9: Comparison between the circular single-sided scanner and the approximated elliptical one with respect to the distance, the gradient and the magnetic field strength H. The side length of the approximated elliptical scanner is 0.03 m for the outer coil and 0.02 m for the inner coil.

		Distance / m	Gradient / T/m	$H / T/\mu_0$
Circular coil	x direction	0.02	-6.378	$-1.679 \cdot 10^{-5}$
	y direction	0.016	0.315	$-6.258 \cdot 10^{-19}$
	z direction	0.016	0.315	$1.546 \cdot 10^{-17}$
Approximated elliptical coil	x direction	0.026	-0.3974	$1.702 \cdot 10^{-4}$
	y direction	0.016	0.2340	$-6.774 \cdot 10^{-18}$
	z direction	0.016	0.1604	$-2.152 \cdot 10^{-18}$

An overview of the gradient strength with regard to the influence of the increasing side length of the approximated elliptical coil is shown in Figure 4.13. When the side length increases, the gradient in x direction becomes flatter. The maximum value decreases, while the minimum value is nearly constant. The gradient in y direction decreases until a side length of 0.06 m for the outer coil and 0.05 m for the inner coil and becomes a bit flatter. Using higher values for the side length such as 0.08 m and 0.07 m or 0.1 m and 0.09 m, the gradient becomes slightly higher again. The gradient in z direction decreases and becomes flatter. From above a certain side length, the gradient is slightly negative, but approximately constant.

Chapter 4 | Experiments and Results

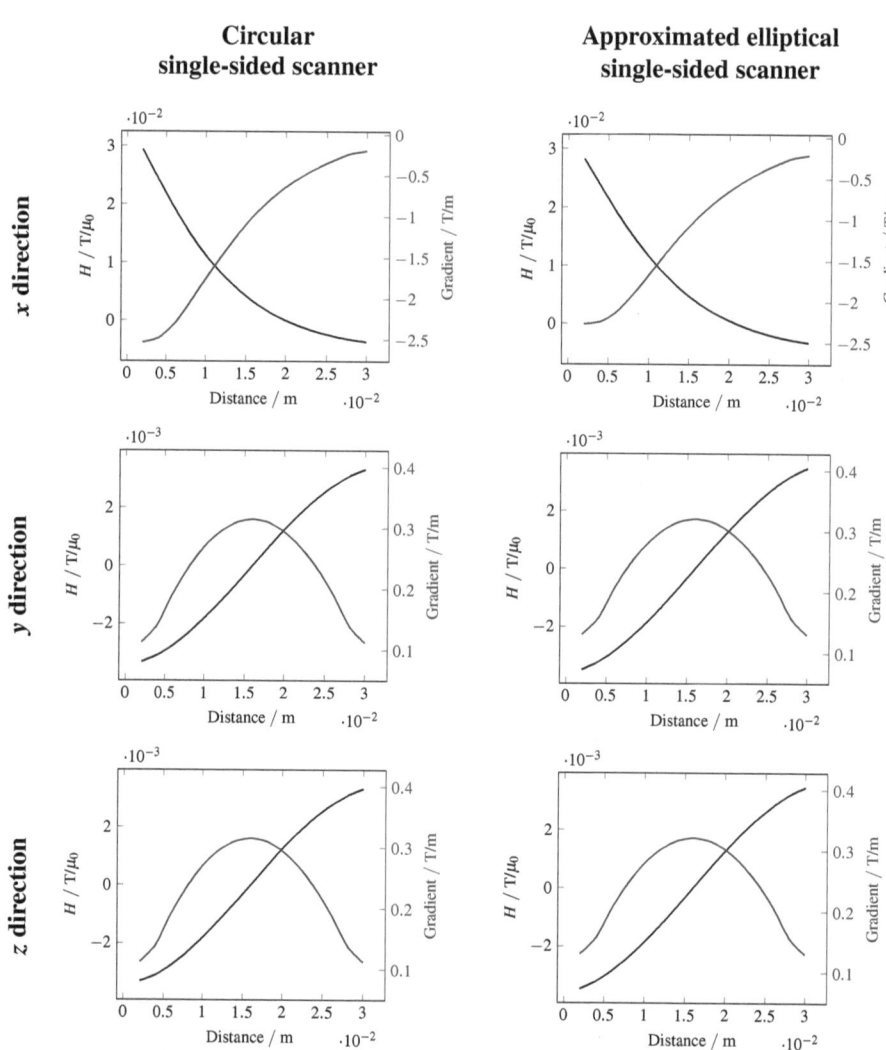

Figure 4.11: Graphical illustration of the magnetic field strength and the gradient of a circular single-sided scanner and an approximated elliptical scanner with a side length of 0 m for the outer and inner coil, considering the x, y and z direction of the magnetic field using a direct current of 55 A for the outer coil and −65 A for the inner coil.

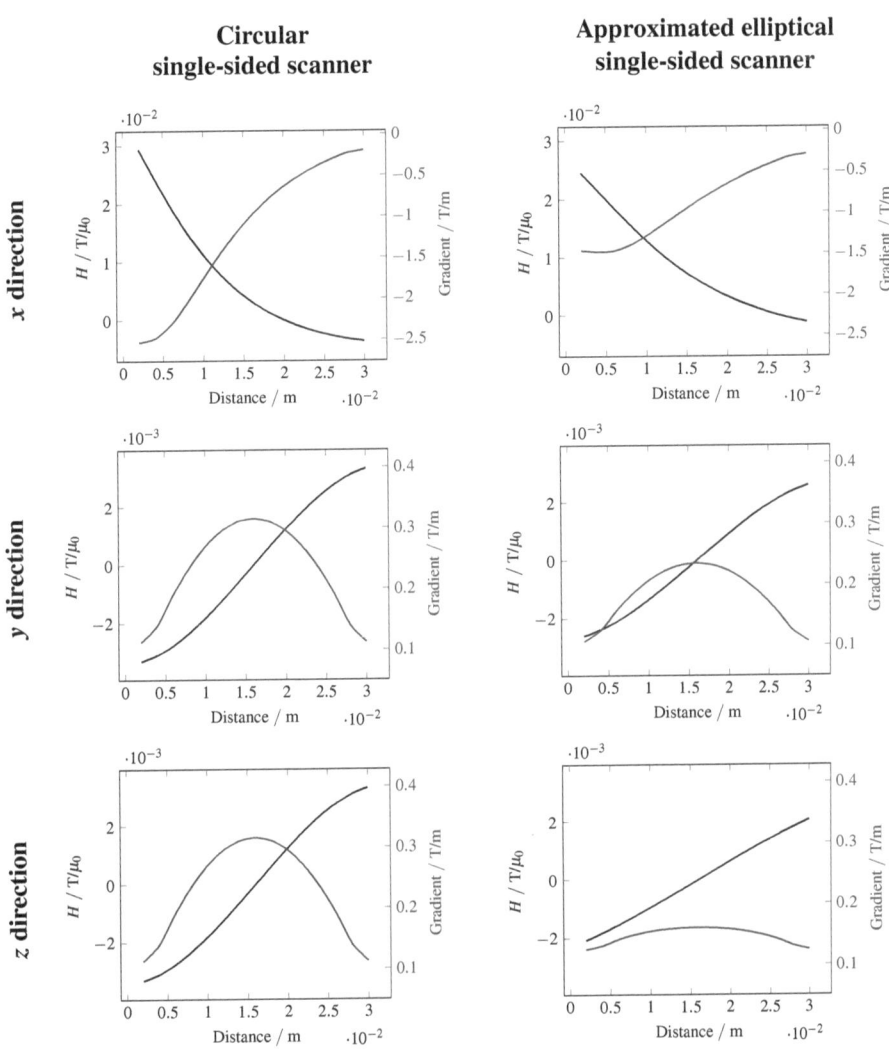

Figure 4.12: Graphical illustration of the magnetic field strength and the gradient of a circular single-sided scanner and an approximated elliptical scanner with a side length of 0.03 m for the outer coil and 0.02 m for the inner coil, considering the x, y and z direction of the magnetic field using a direct current of 55 A for the outer coil and -65 A for the inner coil.

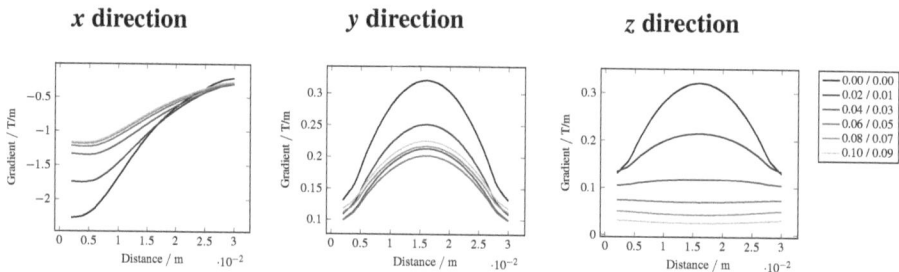

Figure 4.13: Comparison of the gradients in x, y and z direction for different relationships between the side length of the outer coil and the inner coil. The legend entries are given in m, with the first value representing the outer coil and the second value the inner coil.

4.6 Application within an Open MPI Scanner

Similar to the experiments carried out in section 4.5, an open MPI scanner is simulated using approximated elliptical coils. A sample scanner configuration is shown in Figure 4.14. The coils have a distance of 0.1 m to each other, located in the yz plane. The radius of the coils is 0.07 m, the thickness is 0.01625 m and the length is 0.0155 m. The side length of the approximated elliptical coil is varied between 0 m and 0.1 m. A direct current of ± 50 A is used for the experiments. The FOV is centred between the coils with a dimension of 0.06 m × 0.06 m × 0.06 m. The grid used for the calculation is $30 \times 30 \times 30$.

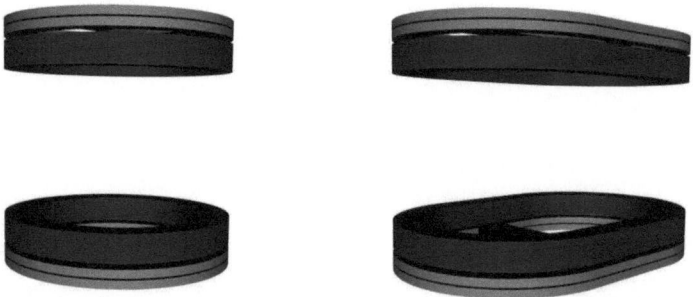

Figure 4.14: Comparative illustration of two open MPI scanner configurations. On the left side is a scanner realised by using circular shaped coils and on the right side is a scanner based on approximated elliptical coils.

4.6.1 Validation

As already described in section 4.5, simulating a circular scanner based on approximated elliptical coils enables a comparison of the magnetic fields to a existing circular shaped open MPI scanner. The comparison of the magnetic fields is performed by using the error metrics described in section 3.5 and results in the following values: $SSD = 8.8673 \cdot 10^{-32}$, $NAD = 6.9523 \cdot 10^{-15}$, $NRMSE = 7.4029 \cdot 10^{-15}$ and $REL = 7.4029 \cdot 10^{-15}$. The resulting values for the NAD, the NRMSE and the REL are all quite near to machine accuracy and the SSD value is even under the machine accuracy. Thus, the difference can be neglected.

For a comparison of the two different scanner geometries, the side length of the approximated elliptical coil is set to 0 m. The results with respect to the magnetic field strength and the gradient strength for the comparison of the two circular scanners are shown in Figure 4.15. The course of the curves for both open MPI scanners are exactly the same. With regard to the magnetic field strength, a point symmetry to zero can be seen for all field components. As shown in Table 4.10, the gradient strength in x direction is the highest, while the values in y and z direction are exactly half of it.

Table 4.10: Resulting values for the magnetic field strength and the gradient strength of an open MPI scanner using ± 50 A. One scanner is based on an existing circular coil, while the other scanner is also a circular shaped scanner based on approximated elliptical coils with a side length of 0.00 m.

		Distance / m	Gradient / T/m	$H / T/\mu_0$
Circular coil	x direction	0.032	0.0113	$1.130 \cdot 10^{-5}$
	y direction	0.032	0.0056	$-5.648 \cdot 10^{-6}$
	z direction	0.032	0.0056	$-5.648 \cdot 10^{-6}$
Approximated elliptical coil	x direction	0.032	0.0113	$1.130 \cdot 10^{-5}$
	y direction	0.032	0.0056	$-5.648 \cdot 10^{-6}$
	z direction	0.032	0.0056	$-5.648 \cdot 10^{-6}$

4.6.2 Extension

The resulting magnetic field strength and gradients for the increased side length of 0.03 m are illustrated in Figure 4.16. The point symmetry of the magnetic field strength as mentioned in section 4.6.1 can be seen in this scanner configuration as well. The values in x and y direction decrease, while the z direction component increases slightly. The values of the gradient strength are listed in Table 4.11. For the components in x and y direction, the gradient strength is lower compared to the circular scanner. With respect to the z direction, the gradient strength is the same.

Table 4.11: Comparison between the circular open MPI scanner and the approximated elliptical one with respect to the distance, the gradient and the magnetic field strength H. The side length of the approximated elliptical scanner is 0.03 m.

		Distance / m	Gradient / T/m	H / μ_0
Circular coil	x direction	0.032	0.0113	$1.130 \cdot 10^{-5}$
	y direction	0.032	0.0056	$-5.648 \cdot 10^{-6}$
	z direction	0.032	0.0056	$-5.648 \cdot 10^{-6}$
Approximated elliptical coil	x direction	0.032	0.0097	$9.693 \cdot 10^{-6}$
	y direction	0.032	0.0040	$-4.017 \cdot 10^{-6}$
	z direction	0.032	0.0056	$-5.673 \cdot 10^{-6}$

The gradient strength of the approximated elliptical coil scanner varies with the chosen side length. An overview of the gradient strength values for a side length between 0 m and 0.1 m is presented in Figure 4.17. With an increasing side length the gradients in x and y direction decrease, while the gradient in z direction slightly increases.

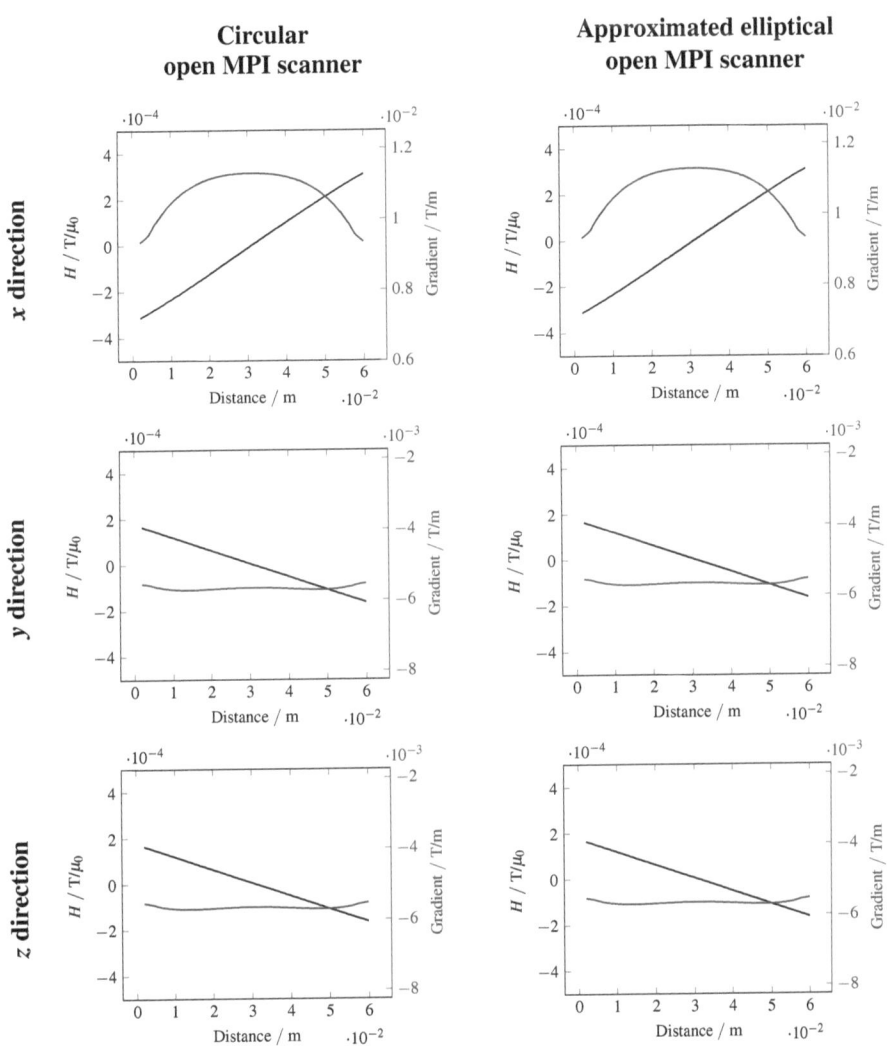

Figure 4.15: Graphical illustration of the magnetic field strength and the gradient of a circular open MPI scanner and an approximated elliptical scanner with a side length of 0 m, considering the x, y and z direction of the magnetic field with a current of ± 50 A.

Chapter 4 | Experiments and Results

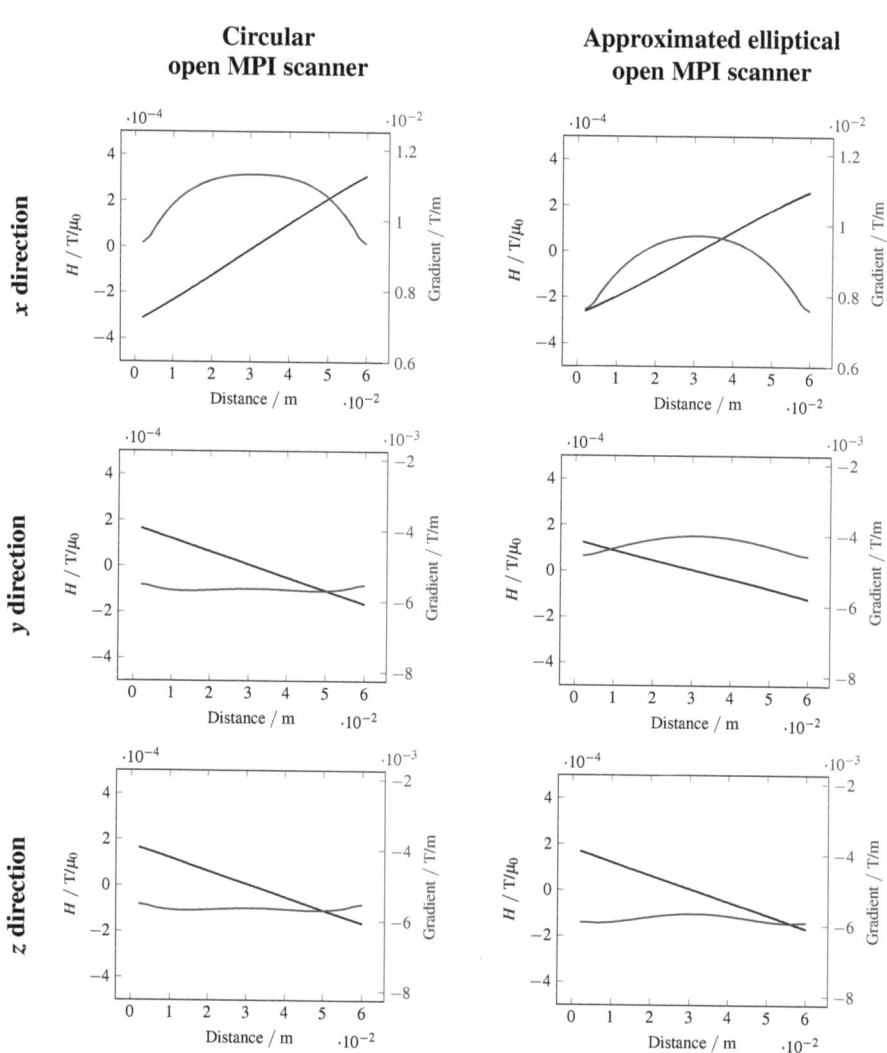

Figure 4.16: Graphical illustration of the magnetic field strength and the gradient of a circular open MPI scanner and an approximated elliptical scanner with a side length of 0.03 m, considering the x, y and z direction of the magnetic field with a current of ± 50 A.

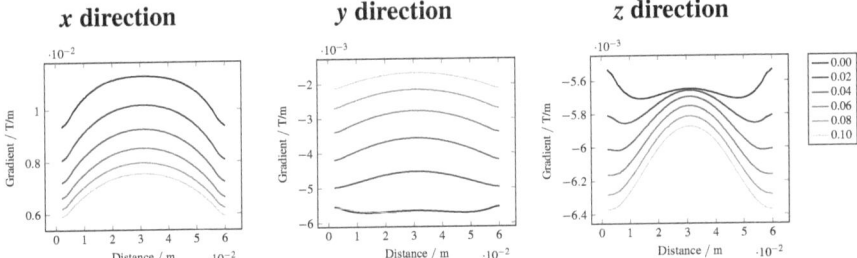

Figure 4.17: Comparison of the gradients in *x*, *y* and *z* direction for different side length values of the approximated elliptical coils. The first value represents the side length of the outer coil and the second value the side length of the inner coil. The legend entries are given in m.

5
Discussion

In this chapter, the results presented previously are evaluated and discussed critically. In accordance with the outline, this chapter is structured similar to the results, beginning with the comparison between the simulation and the measurements and ending with the validation and extension of the simulated approximated elliptical coil.

5.1 Comparison between Simulation and Measurement

Before the actual comparison of the simulation with the built approximated elliptical coil, the results of the noise measurements presented in section 3.4 are interpreted. As shown in Figure 3.16, the frequency distribution of the air measurements can be approximated with a Gaussian probability density function. Through this, the standard deviation of the Gaussian fit can be used to yield statements about the scattering of the values. The calculated standard deviation of this distribution is given by $\sigma = 4.427 \cdot 10^{-6}$ $T\mu_0^{-1}$. Compared to the results of the measurements, the standard deviation is in most cases at least one order of ten smaller than the resulting magnetic field strength values. Therefore, in this work the influence of the noise is neglected.

Chapter 5 | Discussion

For a validation of the approximated elliptical coil implemented to the simulation framework, a comparison with a built approximated elliptical coil is performed. In this context, the considered coil dimensions are the same as the FOV and the direct current used. The current varies between 5 A, 10 A and 15 A to allow an experimental measuring process with simple air cooling. As expected, by doubling and tripling the current strength in the simulation, the values of the magnetic field strength increase accordingly. As shown in Table 4.4, the measured magnetic field values also exhibit the expected course. Due to the fact that the measuring process can be accompanied by displacement problems of the measured coil, such as a slight translation or slant angle to the FOV, the simulated data coincides with the measured data based on rigid transformation. As fast implementable approach to solve this image registration problem, a translation of the FOV in the x, y and z direction is integrated to the comparison. By using the error metrics as introduced in section 3.5, the registration approach is objectively evaluated by comparing the magnetic fields. The resulting values are listed in Table 4.6. Due to the fact that the correction only takes account of a translation, the results are partly about 10 %. If all possible displacement problems were to be considered, the calculated error would decrease. Furthermore, it can be seen in Table 4.6 that the error using the NAD, the NRMSE and the REL decreases with an increased current. This effect is to be expected and is explainable by the stronger magnetic field in combination with the decreased influence of measurement inaccuracies. It is obvious that the difference between simulation and measurement for all field directions and the absolute values is almost zero in the middle of the coil. An interesting effect is that the solder connection and the current carrying conductors can be seen in the REL image of the absolute values in Figure 4.7. This is a possible reason for the differences between simulation and measurement. However, the results of the comparison between simulation and measurement provide promising results and highlight the fact that the implementation of the approximated elliptical coil is representative.

5.2 Validation

Based on the results of the comparison between simulation and measurement, further experiments are performed. In the next steps, the approximated elliptical coil is validated further and an assessment made regarding whether the coil can be used for imaging. The circular shaped coil is selected as the basis for comparing the new coil geometry to other existing coil geometries. This existing coil geometry has already been thoroughly researched and established in MPI. It is used for several simulation studies as well as scanner configurations and is therefore a good choice for validating the approximated elliptical coil. For all of the following experiments, the dimensions of the compared coils or scanners are the same. The only parameter varied during the simulation studies is the side length of the approximated elliptical coil.

Comparing approximated elliptical coils to circular coils for the purposes of validation, the side length of the approximated elliptical coil is set to zero. A good comparison between these two coil geometries can therefore achieved. Regarding the orientation of the FOV, there is one experiment performed with a FOV parallel to the scanner surface and one with a vertical FOV. This is done in order to consider different dimensions with regard to three-dimensional imaging. In both, the FOV do not intersect the coil. Performing an experiment with two circular shaped coils, with one coil based on the approximated elliptical coil, it is expected that there will be no difference between the magnetic fields. The results of the comparison listed in Table 4.7 are in a range between 10^{-7} and 10^{-31}. This means that the values of the error metrics are partly below the machine accuracy, which is $2.2204 \cdot 10^{-16}$ and are therefore negligible.

As mentioned in section 4.5, the application of approximated elliptical coils in a single-sided scanner is evaluated by a comparison to a circular shaped scanner. For the validation, the side length of the approximated elliptical coils is set to zero. The focus of the validation is set to the consideration of the magnetic field strength and the gradient strength. The corresponding courses for these values are illustrated in Figure 4.11. A purely visual comparison of the demonstrated curves suggests no difference between the scanners. The strength of the magnetic field in the x direction decreases as expected with an increased distance to the coil surface. In the y and z direction, the values of the magnetic field strength are symmetrical to zero. These courses are also expected, due to the symmetrical dimensions of the coil geometry used for this scanner. The course of the gradient is also strongly influenced by the chosen coil geometry. At the zero-crossing of the magnetic field strength in the y and z direction, meaning in the middle of the coil, the gradient strength is the highest. The influence of the positive field from the one side and the negative field from the other side cancel each other out at this point, resulting in a strong field difference of the magnetisation and therefore in a high gradient. Similar to the magnetic field strength in the x direction, the corresponding gradient decreases with an increased distance to the scanner. Considering the respective values in the FFP as listed in Table 4.8, the statements given are confirmed. The FFP is at exactly the same position for both scanners with a very similar gradient and magnetic field strength. The small difference occurring in this table could be due to a slightly different discretisation of the coils. This difference could be at the connecting parts of the approximated elliptical coils. However, the difference is small enough to be neglected.

Based on the promising results of the single-sided scanner, it was decided to perform further experiments with another scanner topology, the open MPI scanner. In contrast to the single-sided device, this scanner topology generates a more homogeneous magnetic field between the coils. The validation of the application of approximated elliptical coils is performed in the same way as for the single-sided scanner. First, the magnetic field strength values and the values for the gradient strength are evaluated. Second, the corresponding values in the calculated FFP are considered. Regarding the course of the gradient strength, the expectation is a point-symmetrical course to zero in the x, y and z direction. Due to the symmetrical scanner configuration the highest gradient should be located at the point of the zero-crossing. The calculated values for the magnetic field strength and the gradient strength are illustrated in Figure 4.15. Considering the course of each plot confirms exactly the expectations described previously. In addition to this, the course of the values seems to be very similar based on a visual comparison. This first estimate is confirmed by the FFP values listed in Table 4.10. All of the calculated values, the distance of the FFP, the gradient strength in the FFP and the corresponding magnetic field strength, are nearly equal.

5.3 Extension

With respect to the validation of the application of approximated elliptical coils, the results outlined above are very promising and confirm the implementation of the approximated elliptical coils in a very detailed way. Moreover, these results lead to the extension of the side length of the approximated elliptical coils in order to evaluate a possible advantage or disadvantage compared to circular coils.

Similar to the experimental procedure applied to validate the new coil geometry, the extension of a circular shaped coil to an approximated elliptical coil starts with a comparison of the coils themselves. The focus during this simulation approach is on the maximum and mean values of the magnetic field generated by the coils. The dimensions and parameters as well as the coil positioning are the same as used in the validation procedure. The corresponding results are illustrated in graphic form in Figure 4.8 for the parallel FOV and in Figure 4.9 for the vertical FOV. With respect to the parallel FOV, the results are as expected. The maximum values decrease as the side length increases. It can be seen that the course of the curve starts quite steep and becomes flatter with an increase of the side length. It seems that the values near a limit. This observation may be due to the fact that the influence of the 5 A direct current used decreases while the side length is extended to 1 m. This can also be seen in the potential limits of the magnetisation curve. The field component in the z direction becomes zero, because the coil is extended in this direction, while the other field components approximate a smaller value. The assumptions for the mean values are also confirmed. The values in the x direction decrease, while the values in the y and z direction are nearly constant. The difference between the mean values in the y and z direction is negligible, because the values are below the machine accuracy. With respect to the vertical FOV, the results are quite similar. The difference to the horizontal FOV is the behaviour in the y direction, caused by the orientation of the planar FOV in the xy plane. As a result of this, the mean magnetic field values are not influenced by the extension of the side length, but are constant and equal to the circular ones.

An integration of approximated elliptical coils into a single-sided scanner is the next step of the extension experiments. The side length of the scanner is varied between 0 m and 0.1 m for the outer coil and 0 m and 0.9 m for the inner coil. The remaining parameters and considered field properties are equal to the configuration in the validation process. It should be noted that the whole scanner configuration is based on a circular shaped scanner and is also optimised for this. Therefore, it is conceivable that the results are unable to represent the full potential of the approximated elliptical coils in this scanner topology, but an initial evaluation of the possible application can be obtained. An interesting point here is that the gradient in the direction in which the scanner is not extended, in this case the y direction, is nearly constant as observed by comparing the coils themselves using a vertical FOV. The corresponding results for a sample configuration are illustrated in Figure 4.12. An increase in the side length results in a lower magnetic field strength and a flatter course of the values using approximated elliptical coils. The gradient is also lower. The course of the curves is similar to the curves described during the validation process. It is also demonstrated that the gradient strength in the z direction decreases faster than in the y direction. In order to observe the FFP, the resulting values are listed in Table 4.9. As a result of the extension of the side length, the calculated values differ from the values of the circular shaped scanner. The distance of the FFP to the surface of the coil is increased, while the location in the y and z direction is the same. This means

Chapter 5 | Discussion

that, by increasing the side length, it could be possible to also increase the penetration depth. In this context, the gradient strength is an important factor. The increase of the side length also influences the superimposition of the different field components and in particular the overlay between the magnetic field of the inner and the outer coil is influenced. This leads to a smaller gradient in the x direction. Another interesting point is that the gradients in the y and z direction no longer have the same value. Compared to the gradient in the y direction, the gradient in the z direction is lower. To produce a more well-founded statement regarding this gradient course, the side length is varied, with the results shown in Figure 4.13. The gradient values calculated for the x direction decrease as the side length increases. For the gradient values in y direction too, a decreasing course can be observed, but from a certain side length the values increase again. This effect can be explained by the increased side length and the corresponding increased influence of the longer straight coil part on the field components in the y direction. With respect to the values in the z direction, the gradient strength values decrease to almost zero.

Focussing on the gradient variation, especially in the y direction, the application within the single-sided scanner shows interesting results. The application in an open MPI scanner is very interesting as well. Also of interest is the question of whether the effects observed in the single-sided device can be confirmed in this scanner configuration. Again, the parameters are the same as were used during the validation process. An important detail that should kept in mind is that, in this case, the extension of the side length of the coil is in the direction of the y axis. The results of the application is shown in Figure 4.16. Besides the expected decrease of the magnetic field strength in all considered directions, the course of the gradient confirms the expectations and is similar to the single-sided scanner. The gradient values in the x direction decrease, as do the values in the y direction. Differing from the results observed in the single-sided device, the gradient strength in the z is nearly constant. This effect can also be seen in the values listed in Table 4.11. The position of the FFP stays constantly in the middle between the coils, while the gradient in the x direction decreases slightly. The gradient in the y direction also decreases and the gradient in the z direction is constant. An observation which involves varying the side length to confirm this effect is shown in Figure 4.17. This variation is again performed to support the observation that the variation of the side length influences the gradient. The gradient shows the behaviour as recognised in Figure 4.16: the gradients in the x and y direction decrease, while the gradient in the z direction increases. The gradient effect observed in the single-sided scanner configuration can be confirmed by the results of the open scanner design. A comparison of the course of the curves for the gradient in the y direction in Figure 4.13 and the gradient in the z direction in Figure 4.17 shows an intensification of the gradient course when an open MPI scanner configuration is used.

6

Summary and Outlook

This work deals with the applicability of approximated elliptical coils in MPI. The underlying motivation to introduce such a new coil geometry is based on the idea of integrating a single-sided MPI scanner into a patient table. By using circular shaped coils, a tradeoff between the coil size, the resulting FOV and the patient access has to be taken into account. As an alternative approach to finding a solution for the given situation, the range of existing coil geometries is extended with approximated elliptical coils.

The primary aim of this work is to introduce the new coil geometry and to perform basic evaluations. Therefore, after providing an introduction to the physical principles of MPI and MPI as a tomographic imaging technique, the first part of the work deals with the integration of the approximated elliptical coil geometry into an existing simulation framework. This framework is implemented at the Institute of Medical Engineering at the University of Lübeck and written in C++. It enables the user to choose between different coil geometries and allows them to be arranged to simulate any kind of scanner topology based on the integrated coils. Due to the fact that the implementation in C++ is object orientated, the simulation framework could easily be extended with a new coil.

Chapter 6 | Summary and Outlook

In addition to the integration of the approximated elliptical coil, the framework is extended with a feature to integrate any kind of circular shaped part. On the one hand, this is done to simplify the integration of a new coil geometry including circular parts, while on the other hand this feature is needed to realise approximated elliptical D-coils.

After the extension of the simulation framework, a coil mould is constructed to realise the approximated elliptical coils using litz wire. The construction of the coil mould is performed using the CAD software 'SolidWorks'. Due to its good mechanical properties, high dimensional stability as well as outstanding wear and friction behaviour, the coil mould is made of POM.

A first validation of the simulated approximated elliptical coil is performed by a comparison between simulation and measurement. Due to this, a priori air measurements are performed to exclude the influence of noise on the measured data. The deviations during the measurement process according to a slightly shifted FOV are minimised by a translation of the simulated FOV. The comparative consideration of the coils using certain error metrics shows a good coincidence between the simulated and measured data.

Further, the approximated elliptical coil is compared to a circular shaped coil. This is done to validate the coil and to evaluate the influence of the side length of the coil. By setting the side length of the approximated elliptical coil to $0\,m$ it is shown that the difference between both coils can be neglected. Where the side length of a coil in the yz plane is extended, the magnetic field strength decreases with an increasing distance in x direction, while the fields in y and z direction are nearly constant. In light of these results, which are in line with expectations, further experiments would be useful.

In the next step, the application within a single-sided scanner and an open MPI system is evaluated. As in the previous experiments, the scanner based on the approximated elliptical coil is compared to an equivalent circular scanner for the purposes of a validation. While in the single-sided configuration minimal, negligible differences appear, there is no difference to the open MPI scanner. The extension of the circular scanner to approximated elliptical systems makes it clear that the idea of using such a coil geometry could be a good alternative / advancement to the circular shaped scanners.

This initial feasibility study has yielded highly promising results for the future design of MPI coils. The foundations have been laid for more in-depth and detailed statements regarding the behaviour of approximated elliptical coils. More advanced research, such as the optimisation of the coil geometry with respect to a specific medical application, is therefore absolutely essential. Future research is strongly recommended, since this will without doubt give rise to new and interesting approaches involving approximated elliptical coils. The extension of the simulation environment also implemented, which facilitates the use of circular segments as easy-to-integrate parts of a coil, is not only an excellent solution for applications involving approximated elliptical coils, but also harbours tremendous potential for the design and simulation of new coil geometries.

As an additional project for validating the simulated compared to the measured magnetic fields, a more comprehensive study of the registration problem could be conceived of. By minimising the difference through the rotation or tilting of the coil during measurement, it could be possible to decrease the difference between the fields to a minimum and to determine them as accurate as possible. Regarding the coil geometry itself, one interesting approach would be to integrate an elliptical coil into the framework to analyse the difference between these two coil geometries and to evaluate possible advantages and disadvantages of one of the geometries. For the application within a scanner, it is necessary to use alternating current to simulate the trajectory of the FFP over time. The possibility of a rectangular shaped scan of the FOV is a significant factor with respect to the applicability of the coils and the possible realisation of such a scanner. In relation to this, the variation of the coil dimensions is an interesting field of research for evaluating the different relationships between the parameters and the influence on the magnetic field and especially the gradient strength in the FFP. One interesting approach based on the newly implemented feature involving circular coil parts would be the simulation of rectangular shaped coils with rounded corners. By simulating this modification to the rectangular shaped coils, the difference between the simulation and the coil realised using litz wire would be smaller.

Bibliography

[1] J. W. Anthony. *Handbook of Mineralogy: Halides, hydroxides, oxides,*. Mineral Data Publishing, Tucson, 1997.

[2] K. T. Bae. Intravenous Contrast Medium Administration and Scan Timing at CT: Considerations and Approaches. *Radiology*, 256:32–56, 2010.

[3] S. Biederer. *Magnet-Partikel-Spektrometer: Entwicklung eines Spektrometers zur Analyse superparamagnetischer Eisenoxid-Nanopartikel für Magnetic-Particle-Imaging*. Vieweg und Teubner, Wiesbaden, 2012.

[4] A. A. Bogdanov Jr., M. Lewin, and R. Weissleder. Approaches and agents for imaging the vascular system. *Advanced Drug Delivery Reviews*, 37(1–3):279–293, 1999.

[5] W. F. Brown. Thermal fluctuations of a single-domain particle. *Physical Review*, 130(5):1677–1686, 1955.

[6] T. M. Buzug. *Computed Tomography - From Photon Statistics to Modern Cone-Beam CT*. Springer-Verlag Berlin, Heidelberg, 2008.

[7] C. Caizer. T^2 law for megnetite-based ferrofluids. *Journal of Physics Condens. Matter*, 15:765–776, 2003.

[8] J. M. D. Coey. *Magnetism and Magnetic Materials*. Cambridge University Press, 2010.

[9] L. Elsner, I. Koltracht, and P. Lancaster. Convergence properties of ART and SOR algorithms. *Numerische Mathematik*, 59(1):91–106, 1991.

[10] R. M. Ferguson, K. R. Minard, and K. M. Krishnan. Optimization of nanoparticle core size for magnetic particle imaging. *Journal of Magnetism and Magnetic Materials*, 321(10):1548–1551, 2009.

[11] R. Ferzli and L. J. Karam. A No-Reference Objective Image Sharpness Metric Based on the Notion of Just Noticeable Blur (JNB). IEEE *Transactions on Image Processing*, 18(4):717–728, 2009.

[12] D. Finas, B. Ruhland, K. Baumann, T. Knopp, T. F. Sattel, S. Biederer, K. Lüdtke-Buzug, T. M. Buzug, and K. Diedrich. Sentinal Lymphnode Detection in Breast Cancer by Magnetic Particle Imaging Using Superparamagnetic Nanoparticles. *Magnetic Nanoparticles: Particle Science, Imaging Technology, and Clinical Applications*, B. 1:205–210, 2010.

[13] D. Freedman and P. Diaconis. On the Histogram as a Density Estimator: L_2 Theory. *Probability Theory and Related Fields*, 57(4):453–476, 1981.

[14] D. C. Giancoli. *Physik: Lehr- und Übungsbuch*. Pearson Studium, 2009.

[15] B. Gleich and J. Weizenecker. Tomographic imaging using the nonlinear response of magnetic particles. *Nature*, 435(7046):1214–1217, 2005.

[16] B. Gleich, J. Weizenecker, and J. Borgert. Experimental results on fast 2d-encoded magnetic particle imaging. *Physics in Medicine and Biology*, 53(6):N81–N84, 2008.

[17] R. C. Gonzales and R. E. Woods. *Digital Image Processing*. Pearson Prentice Hall, New Jersey, 2008.

[18] P. Goodwill and S. M. Connolly. The X-Space Formulation of the Magnetic Particle Imaging Process: 1-d Signal, Resolution, Bandwidth, SNR, SAR, and Magnetostimulation. *IEEE Transactions on Medical Imaging*, 29(11):1851–1859, 2010.

[19] P. Goodwill and S. M. Connolly. Multi-Dimensional X-Space Magnetic Particle Imaging. *IEEE Transactions on Medical Imaging*, 30(9):1581–1590, 2011.

[20] P. Goodwill, K. Lu, B. Zheng, and S. M. Connolly. An x-space magnetic particle imaging scanner. *Review of Scientific Instruments*, 83(3):033708–033708–9, 2012.

[21] K. Gräfe, T. F. Sattel, K. Lüdtke-Buzug, D. Finas, J. Borgert, and T. M. Buzug. Magnetic-Particle-Imaging for Sentinel Lymph Node Biopsy in Breast Cancer. *Springer Proceedings in Physics*, 140:237–241, 2012.

[22] F. J. Gravetter and L. B. Wallnau. *Statistics for the Behavioral Sciences*. Cengage Learning, Stamford, 2008.

[23] M. Grüttner, M. Gräser, S. Biederer, T. F. Sattel, H. Wojtczyk, W. Tenner, T. Knopp, B. Gleich, J. Borgert, and T. M. Buzug. 1D-Image Reconstruction for Magnetic Particle Imaging Using a Hybrid System Function. *IEEE Nuclear Science Symposium and Medical Imaging Conference*, pages 2545–2548, 2011.

[24] A. Halkola, T. M. Buzug, J. Rahmer, B. Gleich, and C. Bontus. System Calibration Unit for Magnetic Particle Imaging: Focus Field Based System Function. *Springer Proceedings in Physics 140*, pages 27–31, 2012.

[25] E. H. Hall. On a New Action of the Magnet on Electric Currents. *American Journal of Mathematics*, 2(3):287–292, 1879.

[26] S. Kaczmarz. Angenäherte Auflösung von Systemen linearer Gleichungen. *Bulletin International de l'Académie Polonaise des Sciences et des Lettres*, 35:355–357, 1937.

[27] H. Kaden. *Wirbelströme und Schirmung in der Nachrichtentechnik*. Springer-Verlag Berlin, Heidelberg, 2006.

[28] W. Kaiser. *Kunststoffchemie für Ingenieure*. Carl Hanser Verlag, München, 2007.

[29] D. Kim, S. Parka, J. H. Lee, Y. Y. Jeong, and S. Jon. Antibiofouling Polymer-Coated Gold Nanoparticles as a Contrast Agent for in Vivo X-ray Computed Tomography Imaging. *Journal of the American Chemical Society*, 129(24):7661–7665, 2007.

[30] H. Klingbeil. *Elektromagnetische Feldtheorie*. Vieweg und Teubner, Wiesbaden, 2011.

[31] M. Knobel, W. C. Nunes, L. M. Socolovsky, E. Di Biasi, J. M. Vargas, and J. C. Denardin. Superparamagnetism and Other Magnetic Features in Granular Materials: A Review on Ideal and Real Systems. *Journal of Nanoscience and Nanotechnology*, 8:2836–2857, 2008.

[32] T. Knopp. *Effiziente Rekonstruktion und alternative Spulentopologien für Magnetic-Particle-Imaging*. Vieweg und Teubner, Wiesbaden, 2011.

[33] T. Knopp and T. M. Buzug. *Magnetic Particle Imaging: An Introduction to Imaging Principles and Scanner Instrumentation.* Springer-Verlag Berlin, Heidelberg, 2012.

[34] T. Knopp, S. Biederer, T. F. Sattel, and T. M. Buzug. Singular value analysis for Magnetic Particle Imaging. *IEEE Nuclear Science Symposium Conference*, pages 4525–4529, 2008.

[35] T. Knopp, S. Biederer, T. F. Sattel, J. Rahmer, J. Weizenecker, B. Gleich, J. Borgert, and T. M. Buzug. 2D Model-based reconstruction for magnetic particle imaging. *Medical Physics*, 37(2):485–491, 2010.

[36] T. Knopp, M. Erbe, S. Biederer, T. F. Sattel, and T. M. Buzug. Efficient Generation of a Magnetic Field-Free Line. *Medical Physics*, 37(7):3538–3540, 2010.

[37] T. Knopp, J. Rahmer, T. F. Sattel, S. Biederer, J. Weizenecker, B. Gleich, J. Borgert, and T. M. Buzug. Weighted iterative reconstruction for magnetic particle imaging. *Physics in Medicine and Biology*, 55:1577–1589, 2010.

[38] T. Knopp, T. F. Sattel, S. Biederer, and T. M. Buzug. Field-free line formation in a magnetic field. *Journal of Physics A: Mathematical and Theoretical*, 43(1):9pp, 2010.

[39] T. Knopp, T. F. Sattel, S. Biederer, J. Rahmer, J. Weizenecker, B. Gleich, J. Borgert, and T. M. Buzug. Model-Based Reconstruction for Magnetic Particle Imaging. *IEEE Transactions on Medical Imaging*, 29(1):12–18, 2010.

[40] K. M. Krishnan. Biomedical Nanomagnetics: A Spin Through Possibilities in Imaging, Diagnostics, and Therapy. *IEEE Transactions on Magnetics*, 46(7):2523–2558, 2010.

[41] T. Kühn, A. Bembenek, H. Büchels, T. Decker, J. Dunst, U. Müllerleile, D. L. Munz, H. Ostertag, M. L. Sautter-Bih, H. Schirrmeister, A. H. Tulusan, M. Untch, K. J. Winzer, and C. Wittenkind. Sentinel-Node-Biopsie beim Mammakarzinom. *Der Pathologe*, B. 25:238–244, 2004.

[42] S. Laurent, D. Forge, M. Port, A. Roch, C. Robic, L. Vander Elst, and R. N. Muller. Magnetic Iron Oxide Nanoparticles: Synthesis, Stabilization, Vectorization, Physicochemical Characterizations, and Biological Applications. *Chemical Reviews*, 108(6):2064–2110, 2008.

[43] R. Lawaczeck, H. Bauer, T. Frenzel, M. Hasegawa, Y. Ito, K. Kito, N. Miwa, H. Tsutsui, H. Vogler, and H. J. Weinmann. Magnetic iron oxide particles coated with carboxydextran for parenteral administration and liver contrasting. *Acta Radiologica*, 38:584–597, 1997.

[44] P. Le Callet and D. Barba. Visual features for image quality assessment with reduced reference. IEEE *International Conference on Image Processing*, 1:I – 421–424, 2005.

[45] G. Lehner. *Elektromagnetische Feldtheorie für Ingenieure und Physiker*. Springer-Verlag Berlin, Heidelberg, 2010.

[46] P. Leuchtmann. *Einführung in die elektromagnetische Feldtheorie*. Pearson Studium, 2005.

[47] Q. Li and Z. Wang. Reduced-Reference Image Quality Assessment Using Divisive Normalization-Based Image Representation. IEEE *Journal of Selected Topics in Signal Processing*, 3(2):202–211, 2009.

[48] Z.-P. Lianga and P. C. Lauterbur. *Principles of Magnetic Resonance Imaging: A Signal Processing Perspective*. Wiley, New York, 1999.

[49] K. Lüdtke-Buzug. Magnetische Nanopartikel. *Chemie in unserer Zeit*, 46(1):32–39, 2012.

[50] J. C. Maxwell. On Physical Lines of Force, 1861.

[51] J. C. Maxwell. A Dynamical Theory of the Electromagnetic Field, 1864.

[52] J. C. Maxwell. A Treatise on Electricity and Magnetism. *Clarendon Press, Oxford*, 1873.

[53] K. P. McGee, A. Manduca, J. P. Felmlee, S. J. Riederer, and R. L. Ehman. Image metric-based correction (autocorrection) of motion effects: Analysis of image metrics. *Journal of Magnetic Resonance Imaging*, 11(2):174–181, 2000.

[54] J. Modersitzki. *Numerical Methods for Image Registration*. Oxford University Press, 2003.

[55] J. Modersitzki. *Fair: Flexible Algorithms for Image Registration*. Society for Industrial and Applied Mathematics, 2009.

[56] S. Mornet, S. Vasseur, F. Grasset, P. Veverka, G. Goglio, A. Demourges, J. Portier, E. Pollert, and E. Duguet. Magnetic nanoparticle design for medical applications. *Progress in Solid State Chemistry*, 34:237–247, 2006.

[57] L. Néel. Théorie du traînage magnétique des ferromagnétiques en grains fins avec application aux terres cuites. *Annales de Géophysique*, 5:99–136, 1949.

[58] L. Néel. Some theoretical aspects of rock-magnetism. *Advances in Physics*, 4: 191–243, 1955.

[59] J. A. Nelder and R. Mead. A simplex method for function minimization. *The Computer Journal*, 7(4):308–313, 1965.

[60] J. Rahmer, J. Weizenecker, B. Gleich, and J. Borgert. Signal encoding in magnetic particle imaging: properties of the system function. *BMC Medical Imaging*, 9(4), 2009.

[61] J. Rahmer, B. Gleich, J. Weizenecker, and J. Borgert. 3D Real-Time Magnetic Particle Imaging of Cerebral Blood Flow in Living Mice. *Proceedings of the International Society for Magnetic Resonance in Medicine*, (B. 18):714, 2010.

[62] B. Rantner and G. Fraedrich. Therapiestrategien bei extrakranieller Karotisstenose. *Gefäßchirurgie*, 1:61–72, 2005.

[63] B. Ruhland, K. Baumann, T. Knopp, T. F. Sattel, S. Biederer, K. Lüdtke-Buzug, T. M. Buzug, K. Diedrich, and D. Finas. Magnetic Particle Imaging with Superparamagnetic Nanoparticles for sentinel lymph node detection in breast cancer. *Geburtshilfe und Frauenheilkunde*, B. 69:758, 2009.

[64] T. F. Sattel, T. Knopp, S. Biederer, B. Gleich, J. Weizenecker, J. Borgert, and T. M. Buzug. Single-Sided Device for Magnetic Particle Imaging. *Journal of Physics D: Applied Physics*, 42(1):1–5, 2009.

[65] T. F. Sattel, S. Biederer, T. Knopp, and T. M. Buzug. Magnetic Field Generation For Multi-Dimensional Single-Sided Magnetic Particle Imaging. *Proceedings of the International Society for Magnetic Resonance in Medicine*, B.18:3297, 2010.

[66] T. F. Sattel, T. Knopp, S. Biederer, and T. M. Buzug. Open Coil Arrangement for Interventional Magnetic Particle Imaging. *Proceedings of the International Society for Magnetic Resonance in Medicine*, B. 18:945, 2010.

[67] W. Schroeder, K. Martin, and B. Lorensen. *The Visualization Toolkit: An Object-Oriented Approach to 3D Graphics*. Kitware, 2006.

[68] H.-R. Schwarz and N. Köckler. *Numerische Mathematik*. Vieweg und Teubner, Wiesbaden, 2011.

[69] D. W. Scott. *Multivariate Density Estimation: Theory, Practice, and Visualization Multivariate Density Estimation: Theory, Practice and Visualization*. Wiley, New York, 2009.

[70] H. R. Sheikh, M. F. Sabir, and A.C. Bovik. A Statistical Evaluation of Recent Full Reference Image Quality Assessment Algorithms. *IEEE Transactions on Image Processing*, 15(11):3440–3451, 2006.

[71] A. N. Tikhonov, A. V. Goncharsky, V. V. Stepanov, and A. G. Yagola. *Numerical Methods for the Solution of Ill-Posed Problems*. Kluwer Academic Publishers, 1995.

[72] Z. Wang, A. C. Bovik, H.R. Sheikh, and E. P. Simoncelli. Image quality assessment: From error visibility to structural similarity. IEEE *Transactions on Image Processing*, 13(4):1–14, 2004.

[73] J. Weizenecker, J. Borgert, and B. Gleich. A simulation study on the resolution and sensitivity of magnetic particle imaging. *Physics in Medicine and Biology*, 52(21):6363–6374, 2007.

[74] J. Weizenecker, B. Gleich, and J. Borgert. Magnetic particle imaging using a field free line. *Journal of Physics D: Applied Physics*, 41(10):3pp, 2008.

[75] J. Weizenecker, B. Gleich, J. Rahmer, H. Dahnke, and J. Borgert. Three-dimensional real-time in vivo magnetic particle imaging. *Physics in Medicine and Biology*, 54(5):L1–L10, 2009.

[76] M. N. Wernick and J. N. Aarsvold. *Emission Tomography: The Fundamentals of PET and SPECT*. Academic Press, 2004.

[77] J. Zheng, G. Perkins, A. Kirilova, C. Allen, and D. A. Jaffray. Multimodal Contrast Agent for Combined Computed Tomography and Magnetic Resonance Imaging Applications. *Investigative Radiology*, 41(3):339–348, 2006.

Wir verlegen Ihre wissenschaftlichen Schriften

Bachelor- und Masterarbeiten,
Dissertationen und Habilitationen,
Monografien und Tagungsbände, etc.

Kostenlose Verlegung als Buch mit ISBN-Nummer und Aufnahme in die Deutsche Nationalbibliothek

Hochwertiger Buchdruck in nachhaltiger Produktion (FSC-zertifiziert)

Günstiger Bezug von Autorenexemplaren
Weltweite Präsenz Ihres Werkes bei den großen Händlern: Amazon, Thalia, Hugendubel, Barnes & Noble u.v.m. sowie optional als eBook

www.infinite-science.de/publishing

Infinite Science GmbH
MFC 1 | BioMedTec Wissenschaftscampus
Maria-Goeppert-Str. 1, 23562 Lübeck
book@infinite-science.de